The development of
astronomical thought

Plate III. *OLD AND NEW OBSERVATORIES.* Above: *The Rosse 72-inch reflector, photographed by Patrick Moore in 1967; the tube is still to be seen.* Below: *The new Ondrejov Observatory, near Prague in Czechoslovakia, also photographed by Patrick Moore in 1967. It contains a large modern reflector designed for astrophysical use; the mirror is 79 inches in diameter, so that it is slightly larger than the old Birr telescope.*

The development of astronomical thought

by
Patrick Moore,
O.B.E., D.Sc.(Hon.), F.R.A.S.
Second edition

IAN HENRY PUBLICATIONS
1981

Made and printed in Great Britain by
Lowe & Brydone Printers Limited, Leeds, Yorkshire
for Ian Henry Publications, Ltd.,
38 Parkstone Avenue, Hornchurch, Essex, RM11 3LW

Contents

Acknowledgements

My grateful thanks are due to Lawrence Clarke, who has produced all the line drawings for this book, and to Commander H. R. Hatfield, R.N., for allowing me to use his excellent astronomical photographs.

Second Edition

This new edition has been fully revised, and is, I hope, now complete to 1972. I am most grateful to the publishers for their help and encouragement.

PATRICK MOORE

Selsey, August 1979.

1. The Supreme Importance of Man

We live in a changing world. The Earth of today is vastly different from that of the last century; the tempo of life has speeded up out of all recognition, and our whole outlook has been altered. Generally speaking, men of a hundred years ago did not travel much, and their knowledge of what was happening in the rest of the world was comparatively slight. Nowadays, radio and television have put a very different complexion on matters; the news of, say, an earthquake in Turkey, an aircraft disaster off the Peruvian coast or a political crisis in the Middle East is being discussed in all the world's capitals within a few hours at most.

Quite apart from communications systems, science has revolutionised all our lives. Technologists have taken over, and it seems that computers are in the process of taking over from the technologists. The benefits are immense, but so are the risks, and we must admit that the Earth is a much more dangerous place than it used to be in the period when wars were fought out between strictly-limited professional armies. Two atom bombs have already been dropped upon thriving cities. The third would probably mean the end of civilisation as we know it – a situation that would have been unthinkable in the heyday of our grandfathers.

Modern nations accept science, with all its benefits and all its drawbacks. Indeed, there are indications that the general attitude is becoming somewhat blasé; by now, the launching of a new artificial satellite or Moon-rocket no longer qualifies for the headlines of the national newspapers, and is much more likely to be omitted in favour of some 'scoop' involving a strike-leader or an erring politician. There is, too, a growing feeling of scepticism about the path along which science is

1

leading us. Human intelligence is no longer regarded very highly, and we have departed far from the period when Man was regarded as supreme.

In works of literary criticism, a favourite ploy is to frame some hypothetical question and then discuss the answers that might be given by famous figures of the past – Swift, Voltaire, Goethe, or any one of hundreds of others. The procedure is quite interesting, and it can be applied to astronomy. What, for example, would a scientist of the Classical period have replied if asked: 'How important is mankind?' Leaving aside his exact phraseology, the theme of his answer would have been definite enough. 'Man is all-important,' he would have said. 'Our Earth is the centre of the universe, and all the bodies in the sky have been placed there on our behalf. The Sun exists solely to provide us with light and warmth, while the Moon has been created to give us illumination during the night. The stars are small points which are probably fixed to a crystal sphere, and the so-called wandering stars, or planets, may well be pointers, whose positions affect our lives and destinies and can be interpreted for our benefit. From time to time there are warning signs, such as comets, indicating divine displeasure at our actions; these are sent to show that we must mend our ways. In fact, mankind is the reason for the very existence of the universe.'

Of course, he would have elaborated somewhat according to his nationality. If an Egyptian, he would have claimed Egypt as the hub of the world; if a Chinaman, then China would have had the place of honour. And his knowledge of the behaviour of the celestial bodies would have depended upon the period in which he lived (it is by no means certain, for example, that the early races knew about the wanderings of the planets, though the Greeks most certainly did). But these are details. The significant fact is that the universe was thought to exist for man, and not vice versa.

Put the same question to a leading astronomer of the present time, and the answers will be very different. 'Mankind is of no importance whatsoever. The Earth is a small planet, moving round a dwarf star in an ordinary star-system or galaxy. There must be millions upon millions of other inhabited worlds, and many races whose intelligence is incomparably greater than our own. Our relative importance is much less than that of a single drop of water in the Pacific Ocean.'

This change in attitude is due almost entirely to the growth of our astronomical knowledge. Mysticism and religion have very little to do with it, but it is true that orthodox religion did its best to hamper the progress of thought, and succeeded remarkably well until being finally routed less than 500 years ago. The overall change has taken place in fits and starts, and whether we have yet arrived at a proper evaluation of our status in the universe remains to be seen, though we are entitled to be modestly confident.

Many histories of astronomy have been written, and the present book is not intended to be another, except inasmuch as its whole theme has an historical background. What I propose to do, to the best of my ability, is to show how astronomical thought has evolved over the centuries, and to discuss the wider implications of what we have learned. And the first point to be made is that the ancient ideas, ludicrous though they may appear now, were certainly not irrational when first put forward.

Originally, the Earth was thought to be flat. This was natural enough. Apart from obvious local irregularities such as hills and valleys, the Earth does indeed look flat, and there was no valid reason for primitive man to believe otherwise. Also, the Sun, Moon and stars were thought to travel round the world once a day, and this too was in accord with observation. The stars were seen to be fixed with reference to each other, so that the constellation patterns remained the same for year after year; why, then, should not the stars be fastened to some transparent sphere high above the Earth? The ancients had no means of knowing that the so-called fixed stars are in rapid motion. Nothing of the sort was suspected until much later, and only during the last few centuries have we been able to measure the very small individual or *proper* movements of the stars. To all intents and purposes, the constellations we see today are the same as those which must have been watched by the reindeer-hunters of the Ice Age.

In short, there was every excuse for believing that men lived on a flat Earth lying at rest in the exact centre of the universe. And if occupying this privileged position, it followed that humanity must be of unique importance. So far as could be made out, there was only one Earth, and only one Sun; it would have been too much to expect that early peoples would guess that the other stars are suns in their own right.

It was only with the rise of organized religion that more far-fetched ideas became prevalent. The Egyptians, for instance, thought that the sky was made up of the body of the goddess Nut, while some of the races of the Indian subcontinent held remarkable views according to which the world might be carried on the back of a huge tortoise, or perhaps supported by pillars. And yet these strange ideas did not prevent accurate observations from being made. There is not the slightest doubt, for example, that the famous Great Pyramid in Egypt is astronomically oriented, and there is every reason to admire not only the Pyramid-builders but also the Egyptian astronomers and mathematicians.

All the information available to us indicates that each of the early civilisations believed the world to be of special significance. The Chinese thought that a solar eclipse must be due to a fierce dragon which was doing its best to destroy the Sun and so deprive us of our light and heat – so that the only remedy was to scare the dragon off by making as much noise as possible. There is even a story of two Court Astronomers who were executed in 2136 B.C. for neglecting their duties at such a moment of dire peril. Comets were regarded as menacing by all early peoples, and in fact this fear lasted until quite modern times. There was no thought among the Chinese or the Egyptians that the world might be destroyed by sheer chance. Should any terrible disaster occur, it would be due to deliberate intervention upon the part of revengeful gods, and always there was the certainty that the Earth would be singled out.

It has often been said that in the period before 1000 B.C. or thereabouts, the only 'astronomers' in any sense of the term were those of the famous civilizations such as Egypt, China and Babylonia. However, recent developments may well cause a change of view. It has been suggested that Stonehenge, the famous stone circle in Wiltshire, may have been purely astronomical, and used as a primitive sort of computer for predicting eclipses. When this idea was first put forward, by Gerald Hawkins, it sounded somewhat far-fetched, and came in for strong criticism, but by now calculations seem to have gone a long way towards confirming it, though final proof is still lacking. The building of Stonehenge was begun about 1800 B.C. and was complete by 1300 B.C., so that its period is well before that of Greek philosophy. Incidentally, it is also well before the Druids; there is no connection whatsoever between Stonehenge and the Druid cult.

If Hawkins' theory is right, there were astronomers in England well over a thousand years before the arrival of Julius Cæsar. If so, the same may have been true in other parts of the world. Nothing is known about the astronomical views of the Sonehenge builders, because there is a complete lack of written records, but it does seem that they were prepared to look intelligently at the sky and try to forecast what might be due to happen there. Of course, this is not to imply that they had any idea about the real status of the Earth. No doubt they also regarded it as flat and central.

The practical importance of astronomy was realised at a very early date, and, in particular, calendars were drawn up. Much has been written about the Egyptian calendar and the heliacal rising of Sirius – that is to say, the date when the brilliant star Sirius can first be seen in the morning dawn. The Chinese and Babylonian calendars were also worked out by sheer observation, and it is worth noting, too, that in the New World there were accurate calendars in use among the Maya. (The peak of Maya astronomy came somewhat later, but presumably it was quite independent of either Europe or Asia, and it showed that the observers were highly skilled.) By Classical times the calendar had taken on a very modern aspect. After all, there has been only one modification since

Julius Cæsar ordered the astronomer Sosigenes to put
matters in strict scientific order.

This is all very well, but it did not necessarily mean any
proper understanding of the nature of things. Calendars were
drawn up for practical use, and from this point of view it
did not matter in the least whether the Sun went round the
Earth or whether the Earth were in orbit round the Sun. The
calendar would have been the same in either case. Sosigenes,
in fact, did his work very well indeed, though he certainly
would not have credited that the Sun is more important
cosmically than the Earth.

By the time that Julius Cæsar put his calendar reforms in
hand, Classical astronomy was nearing its end, and it cannot
really be said that the Romans added much to what the
Greeks had found out. It is often tacitly assumed that all the
Greek advances came at much the same time, but this is not
so. The first great philosopher, Thales of Miletus, seems to
have been born about 624 B.C., and the last celebrated
astronomer of the Greek school, Ptolemy, died about A.D. 180.
Therefore, the whole story extended over something like 800
years, or a period equal to that which separates the Moon-
rockets of today from the Crusades. To Ptolemy, Thales of
Miletus was as far back in time as we are from Richard
Cœur de Lion.

Early Greek astronomy was drawn partly from Egypt and
partly from Babylonia. It need not concern us here, except in
passing; for instance, Xenophanes of Colophon (570–470
B.C.?) thought that the stars were made of 'clouds set on fire;
extinguished every day, they are rekindled every night like
coals'. But it was not long before the old idea of a flat Earth
was cast aside, for reasons which were both logical and
accurate. The first great step had been taken.

One vital proof concerned the visibility of stars as seen from
different positions on the Earth. To take just one example,
Canopus, which is brighter than any star in the sky apart from
Sirius, never rises from Greece, but it can be seen from more
southerly lands such as Egypt. The further south one goes, the
higher Canopus rises above the horizon. This sort of behaviour

is only to be expected for a spherical Earth, but cannot be explained on the theory that the world is flat. Then, too, the Earth casts a curved shadow on the Moon during a lunar eclipse (Fig. 1), and this indicates that the Earth's surface is itself curved. The Greeks were quite ready to accept observational evidence of this sort, and Eratosthenes of Cyrene (276–196 B.C.) even measured the circumference of the globe with remarkable accuracy. Strangely enough, the value that he gave was more correct than the figure used by Christopher Columbus during the famous voyage of discovery so many centuries later.

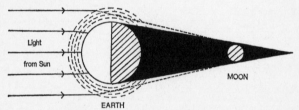

Fig. 1. *Theory of a lunar eclipse. When the Moon passes into the Earth's shadow, its direct supply of sunlight is cut off, but some light is refracted on to its surface by the Earth's atmosphere, so that the Moon does not disappear entirely.*

In the meantime, the movements of the planets were being studied, and star catalogues were being compiled. It had been established that the Moon shines by reflected sunlight, and it was even thought that the stars give off no perceptible heat simply because they are so far away. Eclipses, too, had been explained without recourse to dragons. Generally speaking, scientists were free to teach what they chose. There had been a hint of trouble during the greatest days of Athens, when the philosopher Anaxagoras had been exiled for his impiety in teaching that the Sun must be a red-hot stone larger than the Peloponnesus, but active persecution on religious grounds lay in the future.

Aristarchus of Samos, who lived from 310 to 250 B.C., went much further than Anaxagoras, since he believed that the Sun is larger than the Earth. He also taught that the Earth moves round the Sun, but even the Greeks were not ready for ideas as revolutionary as this, and Aristarchus found few followers. This is often cited as a case of unscientific prejudice, but it must be admitted that Aristarchus could offer no clear-cut proofs, and on the whole it is not really surprising that he was disregarded.

Two points about his work are particularly worth noting. He tried to measure the relative distances of the Moon and Sun, and decided that the Sun must be 19 times the more remote. The real value is almost 400, so that his result was a

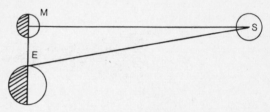

Fig. 2. *Aristarchus' method of measuring the Sun's distance. In the diagram E represents the Earth, M the Moon at the moment of half-phase and S the Sun. The angle at M must be 90 degrees. Aristarchus measured the angle at E and found it to be 87 degrees, so that the angle at S would have to be 3 degrees (for the sake of clarity, the drawing is not to scale). In fact the true angle at E is over 89 degrees, which explained the inaccuracy of Aristarchus' result.*

gross underestimate – and yet in principle, his method was quite sound. He reasoned that since the Moon shines by reflecting the light of the Sun, the position at half-moon must be as shown in Fig. 2, with one angle of the triangle equal to 90 degrees. He measured the second angle, and found it to be 87 degrees, from which it followed that the angle at the Sun must be 3 degrees; the triangle could then be solved. Observation, not theory, was the cause of the trouble. The angle at the Sun is only about 10 minutes of arc, not 3 degrees. Because the Moon has a rough, uneven surface, it is virtually impossible to decide observationally when the phase is exactly half, so that Aristarchus can be forgiven for faulty measurement here, but he also gave the apparent diameter of the Moon as seen from the Earth as 2 degrees, whereas a reasonably careful check would have established that the real figure is only about half a degree.

More than a century later, Hipparchus of Nicæa improved the distance measures considerably. He found that the Sun's distance is more than 1200 times the diameter of the Earth, while the Moon is rather over 30 Earth-diameters away. If we take the accepted Greek value for the Earth's diameter, which was very nearly correct, then the distance of the Sun works out at about 10 000 000 miles and that of the Moon at 260 000 miles or so. Hipparchus was accurate enough for the Moon, and though his distance for the Sun was still much too small he was at least reasoning along the right lines. Yet – and this is the interesting fact – his theory of the universe was completely wrong; unlike Aristarchus, he assumed that the Sun goes round the Earth. Developments in theoretical work and observational accuracy do not necessarily go hand in hand.

Hipparchus was, of course, a brilliant mathematician. He invented trigonometry, he drew up a star catalogue which was much the best of its time, and he discovered the phenomenon known as the precession of the equinoxes (Fig. 3). As it spins, the Earth's axis 'wobbles' slightly, rather in the manner of a top that is running down; the effect is to cause a shift in the position of the celestial pole, and the polar point describes a circle in the sky. It takes 25 800 years to complete a circle of radius 47 degrees, so that the movement is hardly rapid, and Hipparchus' success in detecting it is a tribute to his observational and theoretical skill. Note, incidentally, that when the Egyptian Pyramids were built the polar point lay not near our present Pole Star, but near the much fainter star Thuban in the constellation of the Dragon.

Hipparchus' catalogue of stars was revised by Ptolemy of Alexandria, the last of the great astronomers of antiquity. Ptolemy was undoubtedly a genius, and he was versatile; he drew a map of the known world, compiled by scientific measurement rather than by hopeful guesswork, and as a mathematician and a geographer he was pre-eminent. Most unfairly, he is best remembered now as the leading teacher of a cumbersome scheme according to which the Sun moves round the Earth in a decidedly improbable manner.

Aristarchus' Sun-centred theory had been rejected. The Earth had been replaced in the central position, and it was assumed that all the bodies in the sky must move round it in circles, because a circle is 'perfect' in form and nothing short

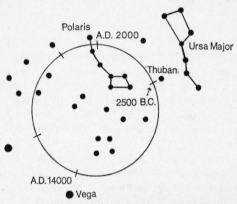

Fig. 3. *The precession of the equinoxes. The celestial pole shifts very gradually, so that the Pole Star we know was not close to the pole in the time of Ancient Egypt. In those days the polar star was Thuban, in the constellation of the Dragon. By A.D. 14000 the nearest bright star to the pole will be Vega.*

of perfection can be allowed in the cosmos. Yet Ptolemy knew quite well that the planets cannot move round the Earth in perfect circles at uniform velocities. Their behaviour is quite different; for instance, Mars, Jupiter and Saturn regularly pause, 'go back' and pause again before resuming their normal eastward motion against the starry background.

Ptolemy therefore supported a system in which the planets moved round the Earth in small circles, or epicycles, the centre of which – the deferents – themselves moved round the Earth in perfect circles. Eventually he had to adopt multiple epicycles, and the result was as artificial as it was unlikely. Yet it fitted the observations, and so it was regarded as satisfactory. It is always known as the Ptolemaic system, though in fact Ptolemy himself did not invent it.

Classical astronomers knew nothing about more modern-type theories. They were completely ignorant of the laws of gravitation; they could not measure cosmical distances with any pretence of accuracy, except with regard to the Moon, and they had no optical aid, so that all their measurements had to be carried out with the naked eye alone. Despite these limitations they made great strides, and they refused to accept any theory that did not agree with the observed results. The strong point about the Ptolemaic system was that it *did* agree, and on this score it could not be faulted.

It is certainly true that even the Greeks were instinctively reluctant to admit that the Earth might not be supreme, and this was an important factor. Yet in view of the state of knowledge at the time, Ptolemy's system of epicycles and deferents appeared just as convincing as Aristarchus' old, revolutionary idea that the Earth moves round the Sun. There was nothing at all irrational in the eventual swing of opinion against Aristarchus, and the decision was made on grounds which were more scientific than mystical. The implicit belief in Man's importance did not stifle the urge to 'find out more'. Attempts at stifling were delayed until much later.

2. Mystics and Doubters

Nothing is known about Ptolemy's personality. His original works have been lost, and have come down to us only by way of their Arabic translations; all we can tell from them is that their author was brilliant by any standards, and periodical attempts by modern scholars to discredit him have not been very convincing. In themselves, his books provide an invaluable summary of the state of scientific knowledge at the end of ancient times. Astronomically, the book usually known by its Arabic title of the *Almagest* is by far the most important.

Ptolemy was a collector of information, though he was also an excellent observer and mathematician in his own right. From our point of view, he lived at the most convenient possible time, because the year of his death, A.D. 180, also marked the end of true Roman greatness. Marcus Aurelius, the philosopher–emperor, died in the same year (actually, we know much more about Marcus Aurelius' career than we do about Ptolemy's), and thenceforth the decline was inexorable, though now and then a comparatively able and energetic ruler gave brief hopes of revival.

Had Ptolemy lived a few centuries earlier, he would have been able to leave us no more than an incomplete account; had he lived much later, at least part of the old knowledge would probably have been forgotten. One particularly disastrous loss was that of the great scientific library at Alexandria, where Ptolemy had lived, and where – 400 years earlier – Eratosthenes had made his calculations about the size of the Earth. It is usually said that the Library was destroyed in A.D. 640 by order of the Arab caliph Omar, who pronounced that either the books opposed the teachings of the Koran and were therefore heretical, or else they supported the Koran and so were superfluous. This may or may not be true, but in any case it

12

seems that much of the Library had been dispersed by the time that Omar came to power. However, the actual method of destruction is irrelevant; the salient point is that the Library ceased to exist.

It is not entirely true to say that astronomy died for more than 600 years after the time of Ptolemy. Very little was done during the period of the collapsing Roman Empire, and progress in China had come to a virtual halt, but a certain amount of observation was carried on elsewhere, notably in India and by the Maya calendar-makers. There is no need to consider this work in our present context, because the attitude was still the same: the Earth lay in the centre of the universe, with Man regarded as supreme. Neither was there any marked change when serious, skilled observing began again, with the Arabs of the 9th century A.D. and later. Yet something must be said about the Arab period, because it involves the old-established link between celestial events and human destinies.

Primitive men had regarded the Sun and Moon as gods, which was not in the least surprising. Later, astrology began to flourish, and continued to do so for a long time; less than 400 years ago it was still classed as a serious science, and even today it retains a considerable grip in some parts of Asia. Astrologers did not suggest that the Sun, Moon or other bodies were actual gods, but they did teach that the positions of the planets in the sky exerted real influence upon mankind, both generally and in individual cases.

Five planets are visible to the naked eye: Mercury, Venus, Mars, Jupiter and Saturn. Adding the Sun and Moon gave a total of seven bodies, and this was regarded as significant, because seven was the mystical number. In astrological practice, the positions of the planets among the stars at the moment of a baby's birth were taken as being all-important, and with this information the astrologer would draw up an elaborate chart, or 'horoscope', so that he could forecast the child's character and destiny. Of course, the horoscope principle could be extended, but the basic principles were always the same.

As early as A.D. 570, Bishop Isidorus of Seville drew a definite distinction between astronomy and astrology, but for many centuries the two remained hopelessly intertwined. Now and then some bold speaker would point out that there can really be no connection between Earthly events and the apparent positions of the planets in the Zodiac; after all, the planets are very much nearer than the stars, and it is absurd to speak of a planet as being 'in' a constellation (quite apart from the fact that the constellations themselves are made up of unconnected stars which simply happen to lie in roughly the same direction as seen from Earth). But among the Arabs, and in mediaeval Europe, astrology was taken very seriously indeed.

By assuming that the planetary positions could affect mankind, the astrologers were in effect claiming that the planets existed for the benefit and guidance of mankind. Indeed, it followed that the stars were there for the same purpose, because astrology could not have been formulated without the constellation patterns. Moreover, the astrologer was bound to believe that the very constellations had been designed for a definite purpose. The Zodiacal groups, such as the Crab and the Lion (Fig. 4), are essential to the cult; the Crab is a somewhat anaemic sign, whereas the Lion is regarded as masculine and vigorous. Of course, the names had been given by the old star-gazers, and Ptolemy's list had included 48 separate groups, including the Zodiacal 12 in which the planets are always to be found. Yet to the astrologer, it was a matter of 'discovering' the names rather than inventing them.

There is no point in saying more here about the astrological cult, except that it underlined the prevalent view that the universe had been fashioned expressly for mankind. Modern astrologers claim the same thing, even though they probably do not realize it. The only real interest of modern astrology is purely psychological.

We cannot be certain how much of Arab astronomy was due to a wish for scientific knowledge. Certainly the horoscope-casters needed to know the positions of the stars and the movements of the planets, and for this information they had to turn to the observers, but it often happened that the skilled observer was himself an astrologer. At all events, the result was beneficial. Ptolemy's great book was translated into Arabic in the year 820, by order of the caliph Al Mamon, and its influence was tremendous. For several centuries the Arab schools of astronomy-cum-astrology flourished, and important research was carried out. For instance, several star catalogues were produced with an accuracy greater than Ptolemy's; the planetary motions were studied, and astronomical tables were greatly improved. There were, of course, major observatories, equipped with instruments used for measuring celestial positions; telescopes still lay in the future,

Fig. 4. *Leo, the Lion. The main star pattern is shown, together with the figure of the mythological lion.*

but naked-eye observations were made with remarkable skill. Incidentally, many of our familiar star-names, such as Betelgeux, Altair and Aldebaran, are Arabic.

The last of the great Arab astronomers was Ulugh Beigh, who established an observatory at Samarkand in 1433. He was a powerful ruler – his grandfather was the Oriental conqueror Tamerlane – and he prepared tables of the planetary motions, drew up an excellent star catalogue and even established an Academy of Science. With his death by assassination, 16 years later, Arab astronomy came to a virtual end. It had improved observational techniques very considerably, but its underlying beliefs were still Ptolemaic even though Ptolemy had been dead for more than 12 centuries.

Moreover, there was an implicit faith in the teachings of Aristotle, often regarded as the greatest of the philosophers of Ancient Greece. Aristotle had said that the Earth is the central body of the universe – and so central it must be; to voice doubts was almost heretical. In any case, there was no obvious need for doubt. The Ptolemaic system accounted quite well for the observed motions of the planets, clumsy though it might be.

This was the general attitude when astronomy began its European revival. Observatories were set up, and after the invention of printing it became possible to distribute star catalogues and astronomical tables on a much wider basis. And yet the overall progress was confined to observing techniques and improved mathematical methods; the few writers who showed signs of really original thought were ignored. One such man was Nikolaus Krebs (Nicholas of Cusa), who lived from 1401 to 1464. In one of his books he even considered the possibility that the Earth might move round the Sun, but no influential scientist was likely to take notice, and Nicholas was either unwilling or unable to support his arguments by positive evidence.

All in all, the theoretical astronomy of mediaeval Europe was no more advanced than that of Classical Greece. So long as the Ptolemaic theory maintained its grip, very little progress could in fact be made. To complicate matters further, there was the influence of the Christian Church, which from a scientific point of view was anything but helpful; the idea that the Earth might not be the hub of the universe was regarded as dangerous, and astronomers were not encouraged to think for themselves. It was only in the 16th century that the status of the Earth became a matter of serious controversy.

3. The Great Dispute

In 1543 Copernicus, a Polish canon, published a book in which he supported the theory that the Earth moves round the Sun. This, then, is the date of the real beginning of the great dispute between the two rival schools of thought. It is almost impossible to decide when the battle ended; the verdict was quite clear by the end of 1609, but the last shots were not fired until more than 50 years later. The *coup de grâce* was given by Newton in 1687, though by that time hostilities were virtually at an end in any case.

The contribution made by Copernicus — or, to give him his proper name Mikołaj Kopernik — is interesting, and has often been misrepresented. He did not invent the heliocentric theory; that had been done in Greek times, with Aristarchus its most famous (if not its first) supporter. Neither did Copernicus solve all the vaious problems with one masterstroke. To be honest, he solved very few of the vital problems, and the system as he left it was no more satisfactory than Ptolemy's had been. In fact, Copernicus made only one real contribution; he dethroned the Earth from its position of eminence in the middle of the planetary system, and put the Sun there instead.

Copernicus was born in 1463. He studied in Poland, then in Italy; later he went back to his home country, where he became Canon of Frombork, practising medicine as well as his official vocation in the Church. His interest in astronomy was lifelong, but he was by no means skilful as an observer, and he concentrated chiefly upon theory. He studied all the ancient works he could find, and decided that none could be regarded as convincing. Something was radically wrong, and he set to work to find out the cause of the trouble.

The Ptolemaic system had become impossibly clumsy. The improved accuracy of observations had shown up ever-increasing discrepancies, and it was certainly not feasible to suppose that the movements of each planet could be explained by a small number of epicycles. Ptolemy had recognized this difficulty, but by Copernicus' time the number of circles required had grown to more than 30, and it was natural for any far-sighted inquirer to cast around for something simpler. No geocentric theory would suffice, but a heliocentric system might be better, and Copernicus adopted it. By 1533 or so he had worked out his scheme, but there were several reasons why he felt reluctant to publish it.

Disbelief and ridicule could be faced; all scientific revolutionaries have had to endure violent criticism. Unfortunately, there was also the Church to be taken into account, particularly as Copernicus was in Holy Orders. It was not likely that Rome would take kindly to a theory that struck so deeply into accepted beliefs, and Copernicus can hardly be blamed for keeping silent. Eventually he did agree to publication, largely because of persuasion from Georg Rhæticus, a Protestant professor of mathematics at the German university of Wittenberg. The book appeared in print during the last days of Copernicus' life, prudently dedicated to Pope Paul III, and with a qualifying preface added by the publisher without the author's permission.

The heliocentric theory, known from that time onward as the Copernican, was received with predictable hostility. For some time few universities dared to teach it, and at first Rhæticus and his colleague at Wittenberg, Erasmus Reinhold, were more or less on their own. The Church was openly hostile, and regarded the whole idea as heretical and revolutionary. Heretical it may have been, but – was it really so revolutionary, after all?

In the sense that it demoted the Earth from its central position, 'yes'; otherwise, most decidedly 'no'. It is truly ironical that after having criticized Ptolemy for being clumsy and artificial in his outlook, Copernicus should fall into precisely the same trap, but this is what happened.

It was clear from the outset that the difficulties about the planetary movements could not be resolved by a straightforward substitution of the central Sun for the central Earth. Copernicus was well aware of this, but he was unable to break free from the old-established notion of perfectly circular orbits, and so his theory was in trouble at once. The only course that suggested itself to him was to bring back epicycles, and as soon as this had been done the system grew in complexity. One example will serve to show what is meant.

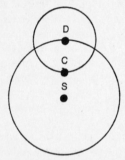

Fig. 5. *Copernicus' theory of the orbit of the Earth. The Earth moved in a circle, the centre of which (C) in turn moved round a point (D) which moved round the Sun (S). Clearly, this system had virtually all the clumsiness of Ptolemy's.*

Figure 5 shows the Copernican view of the Earth's path; the Earth itself was supposed to move in a circle, the centre (C) of which moved round a second centre (D) which in turn moved round the Sun (S). There was nothing in the least simple or straightforward about this, and all the other planets had to be given motions of the same sort, so that the system became just as involved as Ptolemy's had been. Neither were the results noticeably more accurate than those calculated on the Ptolemaic theory. They were no worse, but they were not much better.

Looking back on Copernicus' book, *De Revolutionibus Orbium Cælestium*, it seems at first sight that no progress had been made apart from the one fundamental idea. Moreover, Copernicus gave no conclusive proofs of the correctness of his hypothesis, and firm proof would have been hard to obtain in view of the fact that the hypothesis itself was so faulty. Opponents could well make the same criticisms as those which Copernicus had levelled at Ptolemy: complexity and inconsistency.

It was Erasmus Reinhold who put rather a different complexion upon matters when, in 1551, he completed some new astronomical tables which had been calculated according to Copernican principles. These Prutenic tables, as they were called (in honour of the Duke of Prussia, Reinhold's patron) were better than any of their predecessors, and so they enhanced the prestige of Copernicanism. Actually, the superiority of the tables was due mainly to the fact that Reinhold had been able to use improved observational material, and his results would have been almost as good if he had followed Ptolemy, but this was beside the point. Gradually, more and more astronomers swung over to the Copernican system.

Though Copernicus' book had not appeared until 1543, many people had apparently known about his views as early as 1530, and there had been early criticism from the Church. The first utterance seems to have been to the discredit of Martin Luther, who had said in 1539 that 'this fool wishes to reverse the entire science of astronomy; but sacred Scripture tells us [*Joshua* x. 13] that Joshua commanded the Sun to stand still, and not the Earth'. Luther's colleague, Melanchthon, was equally forthright in 1549: 'It is a want of honesty and decency to assert such notions publicly, and the example is pernicious. It is the part of a good mind to accept the truth as revealed by God and to acquiesce in it.' Protestants were not alone in such views. The Catholic Church soon joined in the battle, and in 1616 *De Revolutionibus* was added to the Papal Index of forbidden books. It remained there until 1835.

The story of the religious objections to Copernicanism has been told many times, but it cannot be ignored here, because it had a profound effect upon scientific thought all through the period between 1543 and the mid-17th century. Church officials were deliberately blind to observational evidence, but astronomers outside Holy Orders were influenced almost as strongly. This is shown by the attitude of Tycho Brahe, whose work led ultimately to the acceptance of the heliocentric theory, but who could not personally bring himself to believe in anything so heretical.

Tycho was born in 1546. He was Danish, and of noble family; his character was strange by any standards, and his career was full of incident. He was astrologically-minded, but he was also an amazingly accurate observer, and he built an observatory on the island of Hven, equipping it with instruments which were the best of their time. Had he possessed telescopes, there is no knowing what he might have been able to achieve. Even with the naked eye, his measures were vastly better than any that had been made before, and his catalogue of stars remained the best for well over a century. He also made excellent measurements of the positions of the planets, particularly Mars.

One event of particular significance occurred in 1572, when a brilliant new star flared up in the constellation of Cassiopeia. At its peak it shone brightly enough to be visible in broad daylight, and Tycho made a long series of observations of it, following it until at last it sank below naked-eye visibility. We know now what it was. It was a supernova – the tremendous outburst of a dying star, resulting in the destruction of the star in its old form (Fig. 6). Nowadays its position is known very accurately, because the débris from the outburst is sending out radio waves which can be picked up with modern equipment. Tycho could not have credited anything of this sort, but at least he realised that the star must be comparatively remote, so that the heavens were not so unchanging as Aristotle had taught.

Obviously, Tycho could not measure the distance of the supernova, but he could show that it must be a long way away, as otherwise it would have shown a shift due to diurnal parallax. The diagram should make this clear (Fig. 7). Let us suppose that the supernova (S) is very close to us, so that it is seen against the background of remote stars. The Earth is rotating, as Tycho knew, so that an observer will be carried from position A to position B over a period of a few hours. From A the supernova will be seen at S^2; from B it will appear at S^1, so that it will show definite movement among the stars in the course of a night's observation. Tycho satisfied himself that this did not happen, and he concluded that the new-

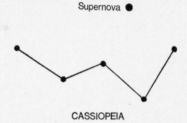

CASSIOPEIA

Fig. 6. *Position of Tycho's supernova of 1572.*

comer must be a genuine star, no nearer than many of the rest. This was it itself a blow to the ancient ideas.

Probably it was the supernova which made Tycho decide to devote his whole life to astronomy. Once settled at his island observatory, he worked unceasingly; every clear night was spent in measuring the positions of the stars and planets, and Hven became very much of a scientific centre. It remained so until 1596, when Tycho left Denmark following disputes with his supporters at Court.

Tycho's observations were good enough to confirm that the old Ptolemaic theory simply did not fit the facts. They were also good enough to prove the truth of the heliocentric idea, but Tycho himself made no efforts in this direction, and it is doubtful whether his mathematical abilities were equal to the task: he was an observer first and foremost. In any case, the 'heresy' of Copernicanism was a fatal obstacle, and as a compromise he developed a system according to which the Sun and Moon moved round the Earth while the planets moved round the Sun. Epicycles and all the other familiar complexities were retained, so that the Tychonic system was a curious sort of hybrid. It appealed to those who had an inward regard for Copernicanism but who shrank from opposing the Church, and it had something of a vogue, but its popularity was brief. Incidentally, it was not basically original. Systems of the same kind had been proposed much earlier, though none had been received with any marked enthusiasm.

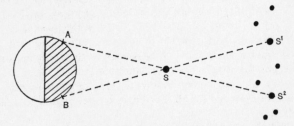

Fig. 7. *Expected diurnal shift of a nearby object. From position A on the Earth, the supernova (S) would be seen at S^2; from position B, the apparent position would be at S^1. Since the supernova of 1572 did not show any such shift, Tycho concluded, quite rightly, that it must be remote.*

There can be no doubt that Tycho's views were coloured by his religious beliefs, but there was scientific evidence as well. He believed, for instance, that the Earth could not be in motion, as a moving world would mean that the stars would show shifts due to diurnal parallax. He calculated that if the shifts were to become too small to be measurable, the star-sphere would have to be at a distance 700 times greater than that between the Sun and the outermost planet, Saturn. He agreed with Ptolemy that the Sun is about 5 000 000 miles from the Earth (Copernicus had given only 2 000 000 miles), and he thought that so immense a gap between the planets and the stars seemed most unlikely.

Tycho's faith in the accuracy of his measures was justified, and his reasoning was logical enough, because he could not be expected to know that the diurnal parallaxes of the stars are indeed too small to be detected. The point to bear in mind is that Tycho's rejection of Copernicanism was based on something more than mere prejudice. And yet the solution to the whole problem was in his hands, had he only known where to look for it.

In his last years, after he had left Denmark and made his home in what is now Czechoslovakia, Tycho had taken as assistant a young German named Johannes Kepler. It was a strange and somewhat uneasy association, since the two men were totally unlike each other; their only common factor was a streak of mysticism, which was hardly surprising in view of the period in which they lived. Whether they would have continued to work together is uncertain, but Tycho died in 1601, and Kepler came into possession of the whole mass of observations that had been accumulated at Hven. At once he set to work in an attempt to solve the problem of the planetary motions once and for all. Fortunately, he concentrated upon Mars.

It had always been known that Mars is further from us than either Mercury or Venus. On the Ptolemaic system it had been placed beyond the orbit of the Sun, while to Tycho it had a path which could take it on the far side of the Earth with respect to the Sun. Its movements had always been particularly hard to explain, and none of the older theories had been entirely satisfactory even when all the numerous epicycles had been taken into account. When Kepler began his labours, he was firmly wedded to the idea of perfect circles, but as the years went by he was forced to the conclusion that no system of this sort could be made to fit in with the data. The agreement was better than anything that had been obtained by Ptolemy or Copernicus, and the discrepancies between theory and observation were so slight that Kepler might have felt justified in disregarding them. The fact that he did not do so was a tribute to his respect for Tycho.

Eventually Kepler found the answer. Instead of moving in a complex orbit made up of compounded circles, Mars travelled in an ellipse, with the Sun at one of the foci. Clearly, the same must be true for all the other planets, including the Earth. This one modification removed the discrepancies; Tycho's measures fell neatly into place, epicycles could be cast aside and the Solar System became straightforward and

orderly. Kepler's First Law was announced in 1609, and with
it came the Second Law, which stated that the radius vector
(i.e. the imaginary line joining the centre of the planet to the
centre of the Sun) sweeps out equal areas in equal times, so
that a planet moves quickest when it is at its closest to the
Sun (Fig. 8). The Third Law, linking a planet's revolution
period with its mean distance from the Sun, followed in
1618.

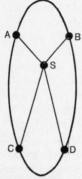

Fig. 8. *Kepler's Second Law. If the planet moves from
A to B in the same time that it takes to move from D
to C, then the sector ABS must be equal to the sector
DSC. (This ellipse is much more eccentric than for any
known planetary orbit – it is however similar to the
orbit of many comets.)*

It is usually said that Kepler proved the truth of the Copernican theory, but this is not entirely true. Kepler did not
verify the Copernican system; he developed his own, which
was much more refined, much more scientific and capable of
hard and fast proof. Admittedly the framework was that of
Copernicus, but the end product was strikingly different.

Not all of Kepler's views were so modern-sounding.
Whether or not he believed in astrology is open to doubt; he
certainly cast horoscopes, but mainly to eke out his meagre
salary as Imperial Mathematician to the Holy Roman
Emperor. On the other hand, he had some extraordinary
ideas, ranging from his strange concept of the 'five regular
solids' to the music of the spheres. He was a curious mixture,

and it is all the more to his credit that he was able to free himself from tradition to the extent that he actually did. His last important work, completed not long before his death in 1630, was the production of new astronomical tables – the Rudolphine tables – which were of course based on his system, and which were so much better than any others that even the opponents of Copernicanism (or Keplerism) were forced to use them.

Logically, the publication of Kepler's Laws ought to have ended the controversy between the geocentric and heliocentric schools. The fact that it did not do so must be put down almost entirely to the influence of the Church. Serious persecution had started, and was demonstrated in tragic manner by the fate of Giordano Bruno, who had been a Dominican friar but had been ostracised by his Order and had subsequently been excommunicated. Bruno was a Copernican, and in particular he believed the stars to be suns, a view that had also been held by some earlier thinkers such as Pietro Manzolli (*circa* 1530) and the English pioneer Thomas Digges, whom Bruno probably met in 1583 during a visit to Britain. Eventually, in 1600, Bruno was condemned to death in Rome, and was burned at the stake. No doubt his scientific opinions were held against him, though it must be stressed that his condemnation was mainly on account of his purely religious heresies; Copernicanism was merely one of his many crimes in the eyes of the Church, and probably not the most grave of them.

Kepler avoided trouble, but his great contemporary Galileo did not. Galileo Galilei had been born in 1564 and was educated at Pisa University, where in 1589 he became professor of mathematics. He was a man of many talents, and was the true founder of experimental mechanics. He was also one of the first to use telescopes for astronomical observation, and in 1609–10 he made a whole series of spectacular discoveries: the mountains and craters of the Moon, the four bright satellites of Jupiter, the phases of Venus, the 'infinite multitude' of stars in the Milky Way and so on. He did not invent the telescope, but he certainly improved it, even though his most powerful instrument was feeble by modern standards. His pioneer work was carried out with a tiny refractor with an object-glass only one inch in diameter.

Galileo had been a Copernican since his early days, and his telescopic discoveries strengthened his beliefs. He pointed out, for example, that the four satellites of Jupiter revolved round their parent planet and not round the Earth, so that at the very least there must be two centres of motion in the Solar System instead of only one. More important still was the behaviour of Venus, which was found to show a cycle of phases from new to crescent, half, three-quarter shape and full. On the Ptolemaic theory, nothing of the kind would have been possible; it had been thought that Venus moved so that it always kept more or less between the Earth and the Sun, so that it could never be seen except in crescent form.

Strictly speaking, the behaviour of Venus was the only phenomenon discovered by Galileo which was in open contradiction to the geocentric hypothesis, though slightly later he observed spots on the solar disc and so put paid to the old-established notion that the Sun must be a perfect body. Yet the telescopic work caused a great mental upheaval, and more and more astronomers found themselves forced to admit that the Earth could not be either stationary or central. The arguments used against Galileo were purely emotional and religious. Some of the Church officials declined a challenge to look through the telescope, which they believed to be bewitched. In a famous sermon delivered in Florence in 1614, a Dominican friar named Tommáso Caccini expressed the wish that all mathematicians should be banished from Christian states as fomenters of heresy. Astronomers, as a class, were not particularly impressed by this suggestion.

Galileo was warned to keep his ideas to himself. Unfortunately for his personal well-being, he was outspoken and impulsive; also he had faith in the integrity of the Church as a whole, and he placed misguided trust in Pope Urban VIII, with whom he had been on good terms before Urban had been elevated to the papacy. When Galileo published his classic

Dialogue, in 1632, the authorities took action. The now-ageing scientist was called to Rome, put on trial and forced to make a public recantation of the false and heretical view that the Earth moves round the Sun. It was a pitiful charade, and Pope Urban emerged from it with no credit whatsoever.

Galileo was not tortured, and after the trial he was allowed to live peacefully under a form of house arrest until his death in 1642. Vindictive to the end, the Pope refused to allow the Tuscan Ambassador to erect a monument over his tomb; to the Church, Galileo was an outcast. Yet there was a major difference between the Roman attitude to Galileo, and Tycho's criticisms of Copernicus. It is true that Tycho regarded Copernicus as heretical, but he rejected the heliocentric theory on scientific grounds also, and to him these grounds seemed perfectly valid (the fact that he was mistaken is beside the point). There was no science in the condemnation of Galileo, and neither was there any vestige of common sense.

It is significant that after the publication of the *Dialogue*, the only astronomers to keep their faith in the Ptolemaic system were those who were also in the priesthood. Dissentients lingered on into the second half of the 17th century; there was, for instance, an Italian Jesuit named Riccioli, who named the lunar craters in honour of famous scientists, taking care to 'fling Copernicus into the Ocean of Storms' and to relegate Galileo to the status of a very obscure and broken-down crater which is of no importance whatsoever.* In Russia the new ideas were slow to take root, and it was not until the career of Mikhail Lomonosov, a century later, that Copernicanism was accepted there.

Yet to most lay astronomers, the Ptolemaic theory was discredited well before 1650. The first great revolution in thought since Greek times had taken place.

* Riccioli's map, based on observations by himself and his pupil Grimaldi, was published in 1651. Tycho, Ptolemy and of course Riccioli himself are allotted very prominent formations. It is true that the crater named after Copernicus is extremely conspicuous, but its site in the Oceanus Procellarum, or Ocean of Storms, was deliberately chosen.

4. The Seventeenth Century

Astronomical thought did not alter very much between the time of Ptolemy and that of Copernicus. During the next 150 years it altered a great deal, and the outlook by 1690 was different in very respect from that of 1540. It is probably true to say that, despite photography, spectroscopy, Einstein, the Palomar reflector, Jodrell Bank and the Apollo missions, there has been no such drastic change since.

There were several reasons for this rapid development. Tycho Brahe, Kepler, Galileo and, above all, Newton were men of exceptional ability; the invention of the telescope came just at the right moment, and in general the intellectual climate was encouraging, even though the Church did everything in its power to preserve the old ideas. Moreover, one step led logically to another, and the talents of the great scientists were complementary: Tycho was the observer, Kepler the pioneer theorist, Galileo the experimenter and Newton the co-ordinator. There was real significance in Newton's famous comment that he had seen further than other men because he had stood on the shoulders of giants.

The telescope had profound influence, not only because it showed features such as the lunar craters and the Jovian satellites, but also because it allowed measures to be made more accurately, With optical aid, a relatively inferior observer could match Tycho's naked-eye results. Of course, the leading observers of the 17th century were anything but inferior (even if none of them possessed quite Tycho's superb ability). To detail their achievements would be out of place here, but something must be said about one particular experiment carried out by G. D. Cassini, an Italian who was called to Paris to direct the new observatory set up there by the King of France.

Kepler had given the distance of the Sun as 14 000 000 miles, but by the mid-17th century it had become clear that this value was still much too small. To improve it, Cassini measured the parallax of the planet Mars. Kepler's Third Law had given a ratio between a planet's sidereal period, or 'year', and its mean distance from the Sun; the period of Mars was known to be 687 Earth-days, and so if its real distance from the Sun were known, that of the Earth could be calculated. By 1672 Cassini had all the information he needed, and he announced that the Earth–Sun distance, or astronomical unit, should be 86 000 000 miles. The fact that this value was seven million miles too small is neither here nor there, because the general order of magnitude proved to be correct, and made astronomers realize that the Solar System is laid out on a vast scale. Stars showed no measureable parallazes, and so presumably were more remote still.

Another investigation of fundamental importance was carried out by the Danish astronomer Ole Rømer, and completed in 1675. Rømer made careful studies of the eclipses of Jupiter's four bright satellites by the shadow of their parent planet, and he found significant differences between theory and observation; the timings were not in accord with calculation, and were either systematically early or systematically late according to the relative orbital positions of Jupiter and the Earth (Fig. 9). Rømer reasoned, quite correctly, that the discrepancies were due to the fact that light travels at a finite speed. Finally he arrived at a value of 186 000 miles per second for the velocity of light, which was remarkably accurate.

This was significant in many ways. Light takes over eight minutes to reach us from the Sun, and so we see the Sun as it used to be more than eight minutes ago. The stars are so remote that their light takes years to reach us, and we can never see them as they are 'now'; beyond the Solar System, our knowledge of the universe must always be very out of date. Rømer's discovery showed this at once, and henceforth no true scientist could claim that the Earth occupied any privileged position.

During the 17th century, astronomy became all-important as a basis for navigation. Of course it had always been valuable, but the new techniques made accurate navigation a real possibility. In particular, there was every prospect of being able to measure longitude by means of relatively simple observations of the Moon's position relative to the 'fixed' stars. It was with this in mind that England's first official observatory was established by order of that much-maligned

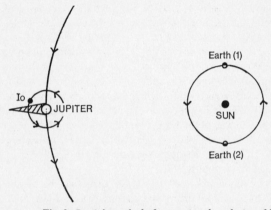

Fig. 9. *Rømer's method of measuring the velocity of light. When the Earth is in position 1, it is approaching Jupiter, and therefore the eclipses of a Jovian satellite such as Io will occur at different intervals than when the Earth is in position 2, and moving away from Jupiter. Jupiter moves comparatively slowly in its orbit, so that for the purpose of this diagram – which is not to scale – it may be regarded as stationary.*

monarch Charles II. Greenwich Observatory was founded in 1675, and the first Astronomer Royal, John Flamsteed, was charged with the task of compiling a star catalogue which would be accurate enough for use by British seamen. After many delays and complications the catalogue was completed, and it was much better than Tycho's. This was not surprising, since Flamsteed had been able to use telescopic sights instead of the relatively primitive instruments that had been employed at Hven.

There is a distinction between improved technical methods and radical changes of view. The alteration in outlook during the latter part of the 17th century was due largely to one man; Isaac Newton, whose classic book of 1787, known generally as the *Principia*, ushered in a new astronomical era. Comparisons are always dangerous, and the common claim that Newton was the greatest mathematician of all time can be challenged, but certainly he has had few equals either before or since.

1687

Newton was born in 1642, the year that Galileo died. He became a student at Cambridge, but his academic career was interrupted by the Plague, and while the University was closed, in 1665–66, he went back to his Lincolnshire home and carried out a series of pioneer investigations into optics and gravitation. He passed sunlight through a prism, and found that the apparently uniform light was split up into a rainbow of colours from red at one end to violet at the other. This may be said to have been the beginning of spectroscopy, even though Newton never carried this particular line of research much further (possibly because the only prisms available to him were of poor-quality glass).

Early telescopes, which collected their light by means of a lens, had suffered from one serious defect. They produced false colour, so that a bright object such as a star would appear to be surrounded by gaudy fringes. Newton realised the cause of the trouble. A lens refracts the different wavelengths unequally, so that the various colours are brought to focus in different places. He could find no remedy, and so he developed a new kind of telescope, using a mirror to collect the light and dispensing with the object-glass altogether. Around 1668 he produced the first reflector, using a metal mirror one inch in diameter. It worked well. Unfortunately, it led Newton into a series of disputes with members of the newly-founded Royal Society, and sparked off the first quarrel with Robert Hooke, who as a theorist and experimenter was second only to Newton himself.

During the Plague period Newton also concentrated upon the movements of the Moon and planets. Most people know the story of how he watched an apple fall from a tree, and reasoned that the force acting on the apple should be the same as the force which keeps the Moon in its orbit. As a matter of fact, the story seems to be true; at any rate, Newton was led on to his fundamental 'inverse square law', though the full mathematical treatment took him many years to complete.

At that period Descartes' theory of vortices was widely favoured. René Descartes, the French philosopher and mathematician, had produced his hypothesis much earlier; it seems to have been complete by 1634, but publication was withheld because of the attitude of the Church, and the full account did not appear until 1664, sixteen years after his death (though it is true that a carefully edited version was issued in 1644). To Descartes, the universe was filled with material, and there was no such thing as a vacuum. Where the material moved, it showed up as stars and planets; the system was one of whirlpool vortices jostling against each other, and when material moved from one vortex to another it was observed in the form of a comet.

The whole idea was brilliant and imaginative, but Descartes made no real attempt to prove it by rigid scientific methods, and it was of course quite unprovable. It was the sort of hypothesis that could be seriously discussed in the middle part of the 17th century, but not afterwards. Again we can appreciate the extent of the tremendous change in outlook over a comparatively few years.

Newton had been taught the vortex theory during his early days at Cambridge, but he could not accept it, and he developed his own ideas. The fact that his researches finally appeared in book form was due largely to the persuasiveness of Edmond Halley, who was a competent mathematician in his own right, but who was modest enought to recognize Newton's superiority. The *Principia* was a true masterpiece. It laid the foundations of dynamical astronomy, set out the three fundamental Laws of Motion, proved the inverse square law and described the concept of universal gravitation upon which so much of later work depends. Newton also discussed the motions of bodies in a resisting medium, and showed mathematically that Descartes' vortex theory simply could not work.

After the *Principia*, astronomy could never revert to its old state, and active persecution by the Church was definitely at an end. Although Copernicus' book remained on the Papal Index for another century and a half, the great dispute was over. Scientific reasoning had triumphed over prejudice, as it was bound to do in the long run. Had Newton not taken the

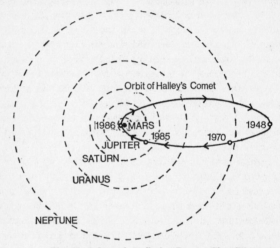

Fig. 10. *Orbit of Halley's Comet. The Comet was at aphelion, beyond Neptune, in 1948. It is now drawing in towards the Sun, and will next come to perihelion in 1986.*

essential steps, someone else would have done so, though probably not so soon.

Halley, who had played so important a part in urging Newton to make his results known, was understandably enthusiastic, and it was he who provided one striking proof of universal gravitation. He had observed a bright comet in the year 1682, and had found that its orbit was significantly similar to those of other comets seen in 1607 and 1531. If the Earth and planets obeyed Newtonian laws, then comets should do likewise. Halley concluded that the three comets were one and the same body, and that the period was about 76 years, so that the next return should take place in 1758. He could not hope to live to see it, but the comet returned as he had predicted; it was first seen on Christmas Night 1758, and came to perihelion in the following spring. Since then, Halley's Comet has returned in 1835 and 1910, and it will come back once more in 1986 (Fig. 10).

There is no need to stress the extent of Newton's influence upon scientific thought. On the other hand, he was not infallible – for instance, his corpuscular theory of light was faulty – and he was also a mystic; he spent much of his spare time in alchemical experiments, and he pored over ancient manuscripts in an attempt to find hidden meanings in them. The mass of notes that he left on these subjects remained unsorted until a few years ago, and it would be pointless to publish them, since they are totally without value.*

Newton died in 1727, but his main scientific work had been carried through in the previous century, and of his major works only the *Opticks* appeared after 1700 (he had delayed its publication until after the death of Robert Hooke, since he knew that otherwise there would be another controversy of the sort that he was anxious to avoid). It is worth noting, too, that Newton was buried in Westminster Abbey with the greatest honour that could be bestowed upon him. Yet less than 100 years before, Pope Urban had forbidden the erection of a monument over Galileo's tomb.

* This does not prevent Newton from being classed as a 'modern'. There is a much more recent parallel in Piazzi Smyth, Astronomer Royal for Scotland, who died as lately as 1900. Smyth was a highly competent astronomer who carried out much valuable work, but he was also obsessed with the pseudo-science of pyramidology, and he believed that the dimensions of the Great Pyramid in Egypt could provide clues to all past and future events in human destiny. Some of his writings are much less rational than anything that Newton wrote about alchemy or kindred subjects.

5. Probing the Universe

The 17th century had been a period of turmoil and revolution. It would have been too much to expect that this sort of rapid progress could be kept up, and after Newton had completed his greatest work there was something of a lull. The next decades marked a period of consolidation and technical improvement rather than radical alteration in outlook, so that while the astronomy of 1700 was completely unlike that of 1600, that of 1800 was merely a logical development of that of 1700.

The Earth had been relegated to the status of an ordinary planet, and it was not even large by planetary standards, since both Jupiter and Saturn were evidently much bigger and more massive. The stars were so far away that their distances could not be measured by 17th- or 18th-century techniques, but it seemed reasonable to assume that the parallax method was sound in theory, so that more accurate measures might be expected to show the stellar shifts. This was the view of John Bradley, who became the third Astronomer Royal (Halley had succeeded Flamsteed, and retained the post until his death in 1742). Bradley was a meticulous observer, and in his younger days, before taking up his appointment at Greenwich, he had set out to see whether he could solve the problem of star distances.

A long baseline would be needed, and Bradley used the diameter of the Earth's orbit, which was known to be about 186 000 000 miles. If a nearby star were observed at six-monthly intervals, it ought to show a parallactic displacement against the background of more remote stars (Fig. 11). Bradley constructed an ingenious device known as a zenith sector, used to measure the apparent positions of stars at or very near the overhead point, and concentrated upon the star

36

Gamma Draconis, which does pass overhead as seen from London. The results were of great interest. Bradley detected a shift, but it was not due to parallax; it was the result of the aberration of light. Despite its importance, Bradley's work provided no new information about the distances of the stars,

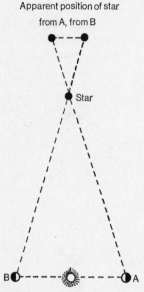

Apparent position of star

from A, from B

Star

B A

Fig. 11. *The parallax of a nearby star. The relatively near star is observed at an interval of six months. When the Earth is at position A, the star will be seen in a different position from that measured six months later, when the Earth has moved round the Sun to position B. The distance AB is known to be 186 million miles (twice the radius of the Earth's orbit). For the sake of clarity, the diagram is very much out of scale.*

and all that could really be said was that they were suns far beyond the limits of the Solar System.

On the other hand, the discovery of aberration provided the first observational proof of the Earth's motion round the Sun (even though there were no longer any doubts upon this score). There is an analogy with the behaviour of raindrops falling on to the umbrella of a man walking through a shower (Fig. 12); to keep dry, the walker must slant his umbrella forward, so that the drops will appear to be falling at an angle instead of coming straight down. With starlight, there is an analogous effect due to the movement of the Earth, which is moving round the Sun at an average velocity of $18\frac{1}{2}$ miles per second, and is changing its direction all the time. The shifts are small, and amount to less than 21 seconds of arc, but Bradley was able to measure them unmistakably. It is not surprising that he failed to detect parallax; even the nearest

Fig. 12. *The effect of aberration. A walker must slope his umbrella forward to keep dry during a shower.*

star shows a parallactic displacement of less than one second of arc, and Gamma Draconis is by no means as close as this. Its parallax is only 0·028 of a second of arc, which was hopelessly beyond the range of Bradley's equipment.

So long as star distances were unknown there was no real prospect of finding out how the star-systems were arranged. Galileo, the first great telescopic astronomer, had seen that the Milky Way is composed of vast numbers of faint stars which look almost as though they were touching one another, but 13th-century astronomers were fairly sure that the stars must be widely spaced even in the most crowded areas of space. It was the aspect of the Milky Way which led Thomas Wright, a Durham teacher, to put forward one of the earliest theories about the shape of the Galaxy. His book on the subject was published in 1750, and is often referred to nowadays as the first major step in cosmological thought.

Actually, this would be an exaggeration. Wright was chiefly a theologian, and he was obsessed with the need for finding suitable positions for heaven and hell. It is quite true that he suggested that the star-system might be disc-shaped, of infinite radius and with the Solar System somewhere near the main plane. Also, he speculated as to whether the dim, misty patches known as nebulae might be outside the Milky Way altogether. For this Wright deserves credit, but it can hardly be said that his theories were based upon strict scientific reasoning, and he seems to have regarded the Sun as something more significant than a run-of-the-mill star. The next important step was not taken until the time of William Herschel, who made his first recorded observation in 1774 and died in 1822.

Herschel was a remarkable man. He was trained as a musician; while still young he came to England from his native Hanover, and spent the rest of his life there. At first, astronomy was no more than a hobby with him, but gradually it absorbed more and more of his time, so that eventually it became his life's work. He taught himself how to make reflecting telescopes, and his greatest reflector had a mirror 48 inches in diameter, though most of his research was carried out with instruments which were smaller and more convenient to handle.

In 1781, while still an unknown amateur, he was examining stars in the constellation Gemini when he discovered an object which showed a small disk. He took it to be a comet, but when its orbit was computed the object turned out to be the planet we now call Uranus. That discovery altered Herschel's whole life. He was made Court Astronomer (not Astronomer Royal, by the way; that post was held by Nevil Maskelyne, founder of the Nautical Almanac), and a modest salary allowed him by King George III of England and Hanover enabled him to give up music as a profession. It is often thought that the discovery of Uranus was

Herschel's greatest achievement, but actually his most important work was in connection with the stars. His Solar System studies were more or less incidental.

Herschel, like Bradley before him, set out to measure star distances. He paid particular attention to double stars, of which there are many in the sky; he reasoned that if one star of a pair lay much closer than the other, as shown in Fig. 13, then there should be a relative shift over six months, due to the

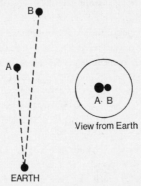

Fig. 13. *An optical double star. Stars A and B are not genuinely associated, since A lies much closer to the Earth, but they appear as a pair. Strangely enough, optical doubles of this kind are in the great minority; most of the double stars known are binary pairs.*

parallax of the nearer star. Like Bradley, he failed – and also like Bradley, he made an unexpected discovery. He found that some of the double stars were genuinely associated 'binary' pairs, with the components moving round their common centre of gravity. Nowadays, studies of binaries make up a very extensive branch of astrophysics.

Herschel's other method was that of star counts. To count every star visible in his telescopes was quite out of the question, and so he concentrated on a relatively few restricted areas, using them as samples. Eventually he came to the conclusion that the Galaxy must be shaped like a 'cloven grindstone', with the Sun near the centre and to all intents and purposes in the main plane. This would certainly account for the aspect of the Milky Way, which proved to be nothing more than a line-of-sight effect; the stars there are not closely packed together, so that appearances are highly deceptive.

Herschel's conclusions were wrong in detail, but correct in outline. The Galaxy is indeed a flattened system, though we now know that the Sun lies a long way out towards the edge (Fig. 14). But even more far-reaching were the tentative suggestions made about some of the nebulae, which Herschel thought might possibly be external systems.

Nebulae had been known for many years; a few of them, notably the Sword of Orion, can be seen with the naked eye.

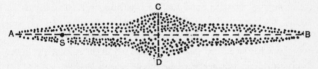

Fig. 14. *Shape of the Galaxy. The diameter AB is now known to be about 100 000 light-years, and the distance CD about 20 000 light-years. The Sun (S) lies well away from the centre of the system. When we look along the direction SA or SB we see many stars in roughly the same line of sight, and it is this which causes the Milky Way effect.*

In 1781, the year in which Herschel discovered Uranus, a catalogue of more that 100 nebulous objects had been compoled by a French observer, Charles Messier, whose hobby was comet-hunting, and who listed the nebulae only so that he could avoid wasting time on them when he found any misty patch which might be a comet. Herschel discovered many more nebulae, as well as clusters of stars, and he realized that not all the objects were of the same variety.

Star clusters were of two kinds: open, such as the Pleiades; and globular, such as the great cluster Messier 13 in Hercules. It was the nebulae which intrigued Herschel. Some of them looked like patches of glowing gas, or what he called 'shining fluid', while others appeared to be made up of stars. The Sword of Orion belonged to the first class, and the Andromeda

Nebula, Messier 31, to the second. Herschel wondered whether the starry nebulae might be systems of the same status as our own Galaxy. He was never at all confident about it, and he seems to have changed his mind more than once, but at least the suggestion had been made, and Herschel's reasons were much more scientific than Wright's had been.

Unfortunately it was out of the question to obtain proof one way or the other. If it were hopeless to measure the parallax of a star with 1800-type equipment, then it was doubly hopeless to attempt to measure that of a starry nebula, which was obviously very remote. Herschel could not solve the problem, and after his death the idea of external galaxies fell into disfavour. It remained out of fashion for many years.

Herschel's greatest contributions lay in the mass of accurate observations that he produced. He is known as 'the father of stellar astronomy', and with ample justification. Every honour that the scientific community could bestow came his way, and his skill and patience as an observer may never have been equalled, but some of his ideas sound curious today; for instance, he was convinced that the Moon and planets were inhabited, and that there might well be intelligent beings living in a cool region beneath the surface of the Sun. It is significant that although few of his contemporaries held such extreme views, he was not ridiculed for his beliefs, as no doubt he would have been had he lived a few decades later. There was a growing feeling that despite the exceptional importance of mankind, thinking beings might well exist elsewhere in the Solar System and beyond.

Although Herschel had not managed to obtain a parallax for any star, there was nothing basically wrong with his method, and in the years following his death fresh attempts were made with more sensitive equipment. As has so often happened in scientific history, success was achieved almost simultaneously by several independent research workers – in this case, three: F. W. Bessel in Germany, Thomas Henderson at the Cape of Good Hope and Otto Struve in Russia. Bessel's announcement came first, and so he is rightly regarded as being the first man to measure the distance of a star. Actually, Henderson had priority so far as the observations were concerned, but ill-health had caused him to delay the reductions.

Bessel had selected the star 61 Cygni, which is dimly visible to the naked eye. It is a wide visual binary, and it has a relatively rapid proper motion, both of which facts indicated that it might be close. Bessel's value of 11 light-years for the distance was remarkably good. Since a light-year (the distance travelled by light in one year) is roughly 6 million million miles,* 61 Cygni lies at around 60 000 000 000 000 miles from the Earth. Alpha Centauri, the star selected by Henderson, is much closer, at only just over 4 light-years; a faint companion of it, Proxima, is still the nearest star known. Struve had the most difficult task, since his star, the brilliant Vega, is over 25 light-years off. His results were understandably less accurate, but they were of the right order.

Stellar parallaxes are by no means easy to measure, even with the aid of modern photographs. For instance, the shift of 61 Cygni over a six-monthly interval is about the same as the apparent diameter of a 1p piece seen from a distance of over 10 miles — and yet 61 Cygni is one of the nearest stars known. With greater distances the parallactic shifts become smaller and smaller, and beyond 250 light-years or so the displacements are swamped in unavoidable errors of observation, so that the method is very limited. All the same, it provides an essential standard, and without it we should still not have a reliable idea of the scale of the Galaxy.

One by-product of this new knowledge was that the Sun was definitely relegated to the status of an ordinary star, by no means exceptional in size or luminosity. To take one example: Sirius, the most conspicuous star in the sky, proved also to be one of the nearest, and to lie at a distance of about $8\frac{1}{2}$ light-years. Its apparent brightness could of course be measured, and fairly straightforward calculations showed that it must be 26 times as powerful as the Sun. Other stars, bright enough to be really prominent but remote enough to show no detectable parallaxes, were clearly more luminous still; Rigel in Orion seems to be about 900 light-years away, and to

* More precisely, 5 880 000 000 000 miles.

outshine the Sun by 60,000 times. The most luminous stars are known to have at least one million Sun-power, while on the other hand the feeble red dwarfs such as Proxima Centauri are very much feebler than the Sun.

So long as measures of star distances were restricted to our own local part of the Galaxy, the overall dimensions and shape of the system could not be determined, and it was a long time before Herschel's picture was improved. However, astronomers tacitly assumed that the Galaxy must be in a state of rotation, with the Sun sharing in the general motion. There was no serious suggestion that the Sun might occupy the central position, with the rest of the stars travelling round it; this would have meant giving ourselves a privileged place, which no longer seemed rational.

Oddly enough there was a popular theory that the centre of the Galaxy was marked by Alcyone, the brightest member of the famous star cluster known officially as the Pleiades and unofficially as the Seven Sisters. This idea seems to have originated with Johann Mädler, who had achieved great (and deserved) acclaim for drawing up the first really good map of the Moon. When Mädler gave up serious lunar charting, in 1840, and became director of the Dorpat Observatory in Estonia, he published various speculative papers, including that concerning the alleged eminence of Alcyone. There seemed to be no good reason to suppose anything of the sort, but the belief persisted for some time. For instance, a well-known astronomical writer named E. B. Denison (afterwards Sir Edmund Beckett) published a book called *Astronomy without Mathematics*, dated 1866, in which he claimed that 'it is almost certain that all the stars we see are in motion, except one, Alcyone . . . which is therefore supposed to be the centre of motion of the visible universe'. By our standards the cosmology of a century ago was still primitive, even though it was no longer obsessed with the need to emphasise our own importance.

6. The Chemistry of the Stars

The telescope had revolutionized astronomy. Less than 180 years separated Galileo's primitive one-inch refractor from Herschel's giant 49-inch reflector, and in the early 19th century there came another important development – the clock drive. If a telescope is being used at high magnification, the field is bound to be small, and a celestial object will move rapidly out of view by virtue of the Earth's rotation. The only remedy is to put the telescope on an equatorial mount, and drive it mechanically so as to compensate for the fact that we are observing from a spinning world. All modern telescopes of any size are mounted in this way, but clock drives were unknown in Herschel's heyday. The first telescope to be equipped with a clock drive was a refractor made by the German optician Joseph von Fraunhofer. It was sent to Dorpat, and was complete by 1817.

Yet, on its own, the telescope is of limited value. It can show surface details on the Moon and planets, but it cannot show a star as anything but a point of light, and without auxiliary equipment we should still know nothing definite about stellar constitution or evolution. The development of the astronomical spectroscope caused a surge of progress comparable with that of the 17th century, though admittedly much less violent. In this field of research, too, Fraunhofer was an outstanding pioneer, and all in all his spectroscopic work was far more important than his clock drive or his exceptional skill in making object-glasses.

Strictly speaking, the story of spectroscopy began with Newton, who carried out the original experiments during the Plague years. He produced the first solar spectrum, but he did little more than note the rainbow effect and interpret it as

45

being due to the compound nature of sunlight. Little more was done until 1802, when an English doctor, W. H. Wollaston, examined the spectrum again and noted some curious dark lines across it. Wollaston had the clue to a far-reaching discovery, but he failed to appreciate its significance, and so the main credit must go to Fraunhofer, who repeated the experiments in 1814 and built the first proper astronomical spectroscope. He used a slit to admit the sunlight, so that the end product was a coloured band from red at the long-wave end to violet at the short-wave end.

Fraunhofer mapped over 500 dark lines in the spectrum of the Sun, and found that their positions and relative intensities never changed. For instance, there were two conspicuous dark lines in the yellow part of the spectrum which were always present, and never varied at all. Laboratory experiments were made, and it was found that various substances gave characteristic spectra; luminous sodium vapour, for example, showed a number of lines, including two which were bright yellow. Fraunhofer apparently wondered whether there might be some connection. Given sufficient time, he might well have found the answer, but unfortunately he died in 1826 at the early age of 39, and for the time being there was nobody of equal ability to follow up his work.

Fraunhofer had also used his spectroscope to examine the light of the stars, and had seen that they too showed spectra which were of the same general type, but the observations were much more difficult and less certain. With the Sun, there is plenty of light available, and so there is no need to use a large telescope to collect every scrap of light for passing into the spectroscope. The intensity of starlight is very feeble by comparison, and so a powerful telescope is essential; moreover, it must be accurately driven.

Fraunhofer's results were published in full, and astronomers took note of them, but there seems to have been no general realisation that a whole new branch of science was waiting to be exploited. (Something of the same attitude was taken much later, between 1930 and 1945, with yet another new study: radio astronomy.) The solar and stellar spectra were regarded as interesting, but the next real advance was delayed for more than 30 years after Fraunhofer's premature death. And at this point let us recall a statement made in 1830 by August Comte, a French philosopher of wide renown. Comte wrote that there were some things that mankind could never hope to find out, and as an example he cited the chemistry of the stars. Man, claimed Comte, could never reach the stars, and so they would always remain mysterious.

It is quite true that at that time very little was known about the stars, except that they were suns. Comte's knowledge was no greater but no less than that of his contemporaries, and his discouraging attitude brought no serious criticism on his head. It was only in 1859 that some classical researches by Gustav Kirchhoff, of Heidelberg University, led to the explanation of the dark Fraunhofer lines in the spectrum of the Sun.

By then, too, much had been heard of the atomic theory proposed by John Dalton, who lived from 1766 to 1844. Dalton had had to face the ridicule which seems to be the lot of every pioneer who departs from strict orthodoxy, but he was responsible for fundamental advances even though his idea of the atom was very different from modern physicists. The essential point he made was that all matter is made up of definite units, and that there are only a relatively few different kinds; nowadays, 92 elements are known to occur naturally. Each element, and each compound, may be expected to yield a characteristic spectrum. The two bright yellow lines of sodium, for instance, cannot be duplicated by any other substance, so that whenever they are seen it may safely be assumed that sodium is present in the luminous source. Kirchhoff's three laws of spectroscopy were linked with this basic idea.

The first law states that an incandescent solid, liquid or gas under high pressure will produce a continuous rainbow spectrum, from red through to violet. The second law lays down that a luminous gas under lower pressure will produce an emission spectrum of disconnected bright lines, and it is here that we come back to Dalton, because each set of lines is related to one particular element or compound. And the third law states that, under suitable conditions, a luminous gas is capable of absorbing light at the wavelengths which it would normally emit.

The Sun has a bright surface, or photosphere, at a temperature of 6000 degrees Centigrade. Surrounding the photosphere is the solar 'atmosphere', where the temperatures are still high but the pressures are much lower (Fig. 15). The photosphere yields a continuous spectrum; the outer gases would normally produce bright lines, but since these lines are less intense

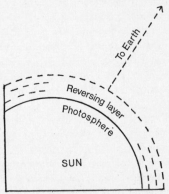

Fig. 15. *Production of the Sun's spectrum. The bright photosphere yields a continuous rainbow spectrum. The solar atmosphere, or reversing layer, would normally produce bright lines, but with the Sun these lines are reversed, and appear dark.*

than the rainbow background they appear in absorption, and show up as dark. (This explanation is deliberately oversimplified, but it will serve to show the general principles.) It follows that the two dark lines in the yellow part of the Sun's spectrum correspond to the two bright yellow lines due to sodium – and therefore there must be sodium in the Sun. Similar identifications can be made for other substances, and by now the total number of elements identified in the solar spectrum is around 70. Naturally, the identifications are not easy, and the whole effect is very complex; iron alone has a spectrum consisting of thousands of lines.

Comte's pessimistic forecast was disproved at one stroke. Since the Sun is an ordinary star, the other stars also can be analysed by means of their spectra, and their 'chemistry' can be studied. Kirchhoff's laws marked the beginning of astrophysics. They also meant a tremendous extension of man's knowledge about the nature of the universe.

The first to take full advantage of the new opportunities were William Huggins in England, and Angelo Secchi in Italy – the first an enthusiastic amateur who set up his observatory outside London, the second a Jesuit priest with full scientific training. Fortunately, their researches were complementary. Secchi examined the spectra of many stars, and fitted them into various definite types, while Huggins concentrated upon a comparative few of the brighter stars and studied them in as much detail as he could. In 1863 Huggins was able to announce the identification of definite elements in the spectra of two bright red stars, Betelgeux in Orion and Aldebaran in Taurus. In the same year, he combined spectroscopy with another promising branch of research, astronomical photography.

The word 'photography' was first proposed by an astronomer: Sir John Herschel, son of Sir William. In its early stages, all the processes of photography were clumsy and inefficient, but from 1839 the methods developed by Louis Daguerre came into wide use, and in March 1840 J. W. Draper, in New York, succeeded in making a Daguerreotype-picture of the Moon, using an exposure of 20 minutes with a 5-inch reflector. The image of the Moon was about an inch in diameter, and the main bright and dark areas on the surface were shown quite clearly.

In an address given to the French Academy of Sciences, François Arago claimed that by this new and exciting method, mapping the Moon would be completed 'in a few minutes'. This was over-optimistic; a really accurate and detailed photographic atlas of the Moon was delayed until the Orbiter space-probes of 1966 and 1967, but at least the possibilities were there. On the other hand, Daguerreotypes were only the beginning, and could not be expected to supersede observation carried out directly at the eye-end of a telescope. The Sun was photographed from 1842, and on 28 July 1851 the German astronomer Berkowski, using the 6¼-inch heliometér at Kónigsberg, obtained a very passable picture of a total solar eclipse, but much research remained to be done before astronomical photography could become really useful.

Total eclipses are not nearly so common as astronomers would wish. For the brief moments when the brilliant disc of the Sun is blotted out by the Moon, the solar surroundings flash into view, and the effect is truly magnificent. It is also important, because there are various investigations which can be carried out only during totality. Even today, solar observers are willing to go on long and difficult expeditions to take advantage of total eclipses, even though there is always the risk of being defeated by bad weather at the critical moment. In modern research, photographic methods are used almost exclusively, but they were of limited practical value in 1851.

Stellar photography was by no means easy in those early days, and it was regarded as a major triumph when Bond and Whipple, using a 15-inch telescope in July 1850, managed to record the image of the bright star Vega. Shortly afterwards they took a photograph of Castor, in Gemini, which is a binary system; the image obtained was decidedly elongated, even though the two components of the pair were not shown separately.* But photography of the stars was slow to progress, partly because of the long exposure-times required and partly because the clock drives then in use were not good enough to keep the star images perfectly steady in the field of view.

Things became much easier with the invention of the wet collodion process, and when the dry plate came into use it was possible to start serious photographic work in stellar

* Castor has a period of 350 years. The components were at their widest, as seen from the Earth, about 1880, and since then the angular separation has decreased; it will continue doing so until the middle of the 21st century, so that Castor is not now the easy pair that it used to be. On the other hand, the binary Gamma Virginis, which is today separable with a small telescope, was difficult in Daguerre's day and will again appear single, except in very large instruments, by the end of the present century.

astronomy. By 1865 Rutherfurd, the American pioneer, was able to take good pictures of the Pleiades and Præsepe clusters in a few minutes, and he concluded that a photographic map of the whole sky was quite practicable. Actually, it was Huggins who first used the dry plate to take star pictures, and in 1863 he obtained the first photograph of a stellar spectrum. It was very different from the clear, complex pictures of today, but it was highly encouraging.

Visual observation of a star's spectrum is beset with difficulties. Everything depends upon accurate measurements of the positions and intensities of the dark lines, and it is obvious that not even the most skilful operator can reach the required standard unless he can obtain a permanent record. Once the spectrum has been photographed, it can be studied at leisure and in comfort with the aid of laboratory equipment. It is probably true to say that, by now, photography has taken over from visual observation, so that the world's largest telescopes are used almost exclusively for photographic work. This did not happen in Huggins' lifetime (he died in 1910), but all the indications were there.

Another development due to Huggins was that of using spectroscopy to measure the velocities of the stars in space. Here he made use of the famous Doppler principle, which had been pointed out in 1842 by Christian Doppler in Austria and extended by the French physicist Armand Fizeau six years later. Anyone with the slightest knowledge of optics is familiar with the basic idea. When a luminous source is approaching, more light-waves per second reach the eye than would be the case if the source were at rest, and the wavelength is apparently shortened, so that the light appears too blue; when the source is receding, the light is reddened. The effect shows up in the positions of the spectral lines as compared with laboratory sources. If the lines are blue-shifted, the star is moving towards us; if the shift is to the red, then the star must be receding – and the amount of the shift is a reliable key to the velocity of approach or recession.

Huggins observed shifts in the spectral lines of some stars, and was able to work out the radial velocities of the stars concerned – that is to say, their towards-or-away motions. Combined with the transverse or proper motions, this gave the real directions and velocities of the stars in space compared with the Sun. For the first time it became possible to study the motions of very remote bodies in the Galaxy, and of course the method was not limited in range. If the spectrum were bright enough to be measured, the red or blue shift, and hence the velocity of the star, could be found.

There was also the question of the nature of the nebulae. It had been definitely established that some of the objects in the Messier and Herschel catalogues were starry, but there was still doubt as to whether the nebulae described by Herschel as giving the impression of 'shining fluid' were starry or not. Huggins settled the matter in 1863 when he was able to examine the spectrum of a planetary nebula in the constellation of Draco. Instead of the familiar rainbow background crossed by dark lines, he found no background, while the isolated lines were bright (emission) instead of dark (absorption). The spectrum was characteristic of a luminous gas at low pressure, and the problem was solved at once. There was a genuine distinction between the resolvable and the irresolvable nebulae, though even Huggins could not appreciate how significant this difference would prove to be.

Meanwhile, Secchi, in Italy, had been busy in classifying the stars into various spectral types. Early in his investigations he found that red stars such as Betelgeux yielded spectra which were not the same as those of yellow or white stars, and it was easy to see that these differences must be due to differences in surface temperature. White or bluish heat is hotter than red heat; thus the white Rigel and the yellow Capella are hotter than orange stars such as Arcturus or red stars of the Betelgeux type. Secchi divided the stars into four classes, but his system was replaced in 1890 by a much more detailed scheme in which the stars are given spectral types denoted by letters of the alphabet. W, O, B and A stars are white or bluish; F and G are yellow, K orange, and M, R, N and S red. The Sun, of course, is a yellow star, and is placed in type G.

As well as providing information about the temperatures,

compositions and radial velocities of the stars, spectroscopy proved able to help in determining their actual luminosities with at least reasonable accuracy. It was realized that a very luminous giant star has a spectrum subtly different from that of a dwarf with the same surface temperature. For instance, both Betelgeux and Proxima are of type M, but Betelgeux is a giant far larger and more luminous than the Sun, while Proxima is very dim by stellar standards – and their spectra are easily distinguished.

More important still was the fact that the intensities of various spectral lines were found to be associated with the actual luminosities of the stars concerned. Consider two stars, A and B, of which the distance of A is already known by parallax methods, while B is more remote and shows no detectable parallactic shift. If the spectra show that B is, say, twice as luminous as A, then its distance can be worked out. Rather confusingly, this is termed 'spectroscopic parallax' – a bad term, as no shifts in position are involved.

One further example of the value of spectroscopy in stellar research is worth giving. In 1889 H. C. Vogel, Director of the Potsdam Observatory in what is now East Germany, was making a careful study of the star Mizar, or Zeta Ursæ Majoris, in the 'tail' of the Great Bear, when he discovered an interesting phenomenon. Mizar is easy to recognize, because it has a fainter star, Alcor, close beside it; in any modest telescope Mizar itself is seen to be made up of two components, one rather brighter than the other, so that it is a wide, easy binary. Vogel concentrated upon the brighter member, Mizar A. He found that periodically the spectral lines became double, only to revert subsequently to their single appearance. Vogel reasoned, quite correctly, that Mizar A was a very close binary, so that the components appeared as one star even when observed through a powerful instrument. The periodical doubling of the spectral lines was due to the Doppler effect.

In Fig. 16, A and A' are the two components, moving round their common centre of gravity; E is the direction of the Earth. In the first position, A is approaching the Earth and A' is receding, so that the Doppler shifts are to the blue and the red respectively, and the spectral lines will be separated. In the second position, there are opposite effects; A is now receding and A′ approaching, so that once more there is a separation of the lines. Midway between these two extremes, the lines will appear as single. (All these effects are of course superimposed on the general shift due to the motion of the whole Mizar system relative to the Earth, but this is known, and can be allowed for.) Vogel had proved the existence of spectroscopic binaries, of which vast numbers are now known.

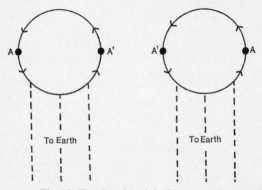

Fig. 16. *The Doppler shift for a spectroscopic binary.* Left: *A shows a blue shift, A′ a red shift.* Right: *A shows a red shift, A′ a blue shift. In each case the spectral lines of the combined system appear doubled.*

The situation had certainly altered since Comte's announcement of 60 years before. The chemistry of the stars was no longer unknown, and astronomers were learning how to probe further and further into the Galaxy. The change had been dramatic indeed; only two centuries elapsed between Newton's pioneer experiments with simple prisms, and Huggins' detection of chemical elements in the stars.

It had become glaringly obvious that the Sun is a normal star, and this was of special importance in the new science of astrophysics, because the Sun is the only star close enough to be examined in real detail. Visual work had begun with Galileo and the other early telescope-users, and the behaviour of the curious dark patches known as sunspots had been followed. It had been found that the Sun shows an 11-year cycle of activity, and it was known that the spots are areas where the temperatures are appreciably lower than those of the surrounding photosphere (4000° C as against the general 6000° C). Yet by itself, the telescope could do no more than show the spots and the associated bright patches known as faculae.

Spectroscopy led to rapid advances in knowledge, and by 1878 more than 30 different elements had been identified in the solar spectrum. In this research, one of the leaders was Joseph Norman Lockyer, of Britain. Lockyer had been on an expedition to observe the total eclipse of 1868, and had found a bright yellow line in the spectrum of the prominences – those magnificent phenomena previously known as Red Flames, visible with the naked eye only during totality. Lockyer's yellow line did not quite agree in position with the two famous sodium lines, and he attributed it to an unknown element which he christened helium. More than a quarter of a century later, in 1895, Ramsay in England, and Cleve and Langlet in Sweden, independently discovered helium in the spectrum of a terrestrial mineral which had been named cleveite in honour of Cleve. We now know that apart from hydrogen, helium is the most plentiful element in the whole universe.

The 1868 eclipse was notable for another significant piece of research. Both Lockyer and the French astrophysicist Jules Janssen had been observing the prominences, now known to be masses of glowing hydrogen above the Sun's surface. It occurred to Janssen that the spectral lines of the prominences were so bright that they might be visible in his equipment without waiting for another eclipse. On the next day he placed the slit of his spectroscope tangential to the Sun's image, and found that the prominence lines could be seen; before long, he was able to devise a method of studying the prominences at any time. Lockyer, in England, had hit upon the same idea; and had actually observed the prominences before he heard of Janssen's success.

Since those days, improved equipment has led to a much better understanding of the Sun's make-up. The spectroheliograph and the monochromatic filter are used to isolate the sunlight emitted by various elements, so that, for instance, the distribution of calcium or hydrogen can be seen on its own. Maps of the solar spectrum are highly detailed, and the number of recognized elements has been more than doubled. Everything that has been learned reinforces the view that there is nothing exceptional about the Sun, and it may not even be unusual in being surrounded by a system of planets.

The idea that the Sun might be a huge globe 'on fire' did not seem to fit the facts, but it was a long time before any acceptable theory could be put forward to explain the tremendous amount of energy streaming out day after day, year after year, century after century. The true answer was not found until the eve of the Second World War. By then there had been new information, too, about the age of the Earth and of the Sun; Man had been shown to be a relative newcomer to the scene.

Plate I. *RADIO AND OPTICAL.* Left: *The Jodrell Bank 250-foot radio telescope.* Below: *Dome of the 24-inch refractor at the Lowell Observatory, Flagstaff, Arizona. (Photographs by Patrick Moore, 1964 and 1967.)*

Plate II (a)

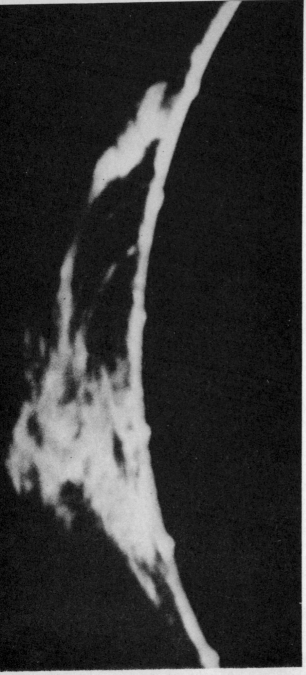

Plate II (b)

Plate II. *ECLIPSES OF THE SUN: OLD AND NEW*. (a) Stonehenge, believed by many authorities to be a primitive computer used for forecasting eclipses. (Photograph by Patrick Moore, 1966.) (b) Solar prominence, 18 Aug. 1947, photographed in calcium light. The prominences can be seen with the naked eye only during a total eclipse, though with modern instruments they can be studied at any time. (Courtesy of Palomar and Mount Wilson Observatories.)

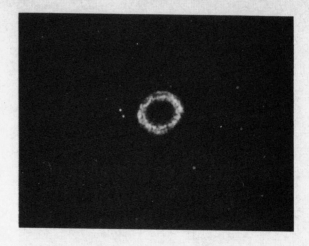

Plate IV. *THE RING NEBULA IN LYRA, a typical planetary.* Above: *As drawn by the third Earl of Rosse in 1851, with the 72-inch Birr reflector.* Below: *Photographed with the Palomar 200-inch reflector.*

Plate V. THE WHIRLPOOL GALAXY IN CANES VENATICI (THE HUNTING DOGS). Left: As drawn by the third Earl of Rosse in 1847, using the Birr reflector. Right: Modern photograph, reproduced by courtesy of Mount Wilson and Palomar Observatories.

Plates VI and VII. *THE ANDROMEDA SPIRAL.*

Plate VI (below): *Photographed by Commander H. R. Hatfield, R.N., with an ordinary camera attached to a driven telescope. Exposure 10 minutes; 16 Aug. 1967. The Spiral can be seen to the upper right, but the structure is not, of course, shown, since the camera was being used 'on its own'.*

Plate VII (right): *The Spiral as photographed with the Mount Wilson 100-inch reflector. (Courtesy of Mount Wilson and Palomar Observatories.)*

7. The Time-Scale of the Universe

The great discoveries of the early 17th century showed that the universe must be far larger than had previously been thought. Kepler's estimate of 14 000 000 miles for the Earth–Sun distance was very much too low, but it was a notable advance on earlier estimates, and it was clear that the stars lay at distances which were tremendous by any standards. Logically, then, the universe ought to be extensive in time as well as in space. This may have been the scientific view, but it was not the attitude of the Church. Between 1650 and 1654 Archbishop Ussher of Armagh, one of the leading ecclesiastical scholars of the period, published some research in which he 'proved' that the Earth came into existence at ten o'clock in the morning of 25 October 4004 B.C. It is hardly necessary to add that Ussher's calculations were entirely Biblical (they consisted chiefly of adding up the ages of the patriarchs) and had no scientific background whatsoever, but they were tacitly accepted by all the religious authorities.

The first definite proofs of the great age of the Earth came with studies of fossils. Many of these fossils were obviously a great deal older than 4004 B.C., and dated back for tens of thousands of years at least. This was something that most Churchmen came to accept, provided, of course, that they could reconcile it with the account of the creation of Man as given in the Book of Genesis.

Religious qualms did not prevent 19th-century research into the Earth's age, but the problem was not easy to tackle. It seemed that the best method might be to concentrate upon the age of the Sun and then 'work forwards', so to speak, since the Sun must be at least as old as the Earth and is probably rather older. This was the view held a century ago, and it is still valid today, but our whole idea about the time-scale has changed.

Originally the Sun was thought to be burning. But a Sun made up of coal, sending out as much energy as the real Sun actually does, would last for less than 10 000 years – long enough to satisfy Archbishop Ussher, but certainly not long enough to meet the needs of 19th-century geologists. Another theory was that the Sun might be kept hot by meteoric infall, but this was soon found to be hopelessly inadequate. The German physicist Helmholtz suggested that gravitation might be the source of solar energy; a shrinking Sun would un-doubtedly radiate, and the contraction would be slow enough to be quite undetectable. Lord Kelvin, the great British physicist, worked out that a decrease of about eight yards per year in the Sun's diameter would be sufficient. Kelvin also thought that the Sun must be cooling down; in his own words, 'cooling and condensation go together'.

Kelvin's reputation was second to none, and remained so up to the time of his death in 1907. He was understandably cautious when discussing the time-scale of the universe, but he did not consider that the Sun could be more than a few million years old, and he regarded 12 million years as a good estimate. On the contraction hypothesis, the shrinkage would continue until the solar diameter had been reduced to half its present value; by then the density would have become as great as that of lead, and the contraction would come to a stop 'through overcrowding of the molecules'.

When calculating the age of the Earth, Kelvin started by assuming that it had originally been as hot as the Sun, and he then worked out the time it must have taken to cool down to its present temperature. He arrived at a figure of between 10 and 100 million years, probably nearer to the lower limit than the higher. Various other lines of investigation led him to the same conclusion, and according to the astronomical evidence it seemed that the Earth could not be very much more than 10 million years old.

Geologists were not in the least satisfied with a time-scale of this sort, and developments following upon the discovery of radioactivity showed that something must be badly wrong. The age of the Earth was re-measured by the decay of uranium into uranium-lead, and the value obtained ran into thousands of millions of years. The latest radioactive determination, from the decay of rubidium, gives the age of the oldest terrestrial rocks as 4500 million years, so that the Earth itself seems to be around 4700 million years old. Whether this value is precisely accurate need not concern us at the moment. It is definitely of the right order, and it at once disposes of the theory that the energy of the Sun (and the stars) comes from nothing more than contraction under the force of gravity. An extreme upper age-limit for the Sun on the contraction hypothesis would be 50 million years, which is not nearly enough.

Meanwhile there had been considerable progress in stellar spectroscopy, and some remarkably significant facts had emerged. It was found, for instance, that red stars were divided into two very distinct classes, giants and dwarfs, with the giants far larger and more luminous than the Sun, and the dwarfs much smaller and dimmer. The giant and dwarf division was also pronounced for the orange and the yellow stars, though with decreasing range, and for the white and bluish stars there was no such separation. E. J. Hertzsprung, of Denmark, and H. N. Russell, of the United States, independently developed what are now known as Hertzsprung-Russell or H-R diagrams (Fig. 17) in which spectral type is plotted against luminosity (or against surface temperature). The giant branch shows up well, and there is a striking tendency for stars to fall along the so-called Main Sequence, from hot stars of early spectral type through to the feeble red dwarfs.

According to a theory due to Lockyer, a star would begin its career by condensing out of the gaseous material in a nebula, and would pass down the giant branch of the H-R diagram, heating up as it shrank. After joining the Main Sequence at type O, B or A, it would pass down until ending its luminous career as a red dwarf of type M. Everything seemed perfectly logical – except for the discrepancy in the time-scale.

Russell, among others, recognized this difficulty, and tried to overcome it by means of his annihilation theory, first put forward in 1913. It was known that of the component particles of an atom, protons are positively charged and electrons negatively charged. Suppose that a proton and an electron could collide; what would be the result? A positive and a negative charge would presumably cancel each other out; $+1 - 1 = 0$, and the particles would suffer mutual annihilation, with the release of a certain amount of energy.

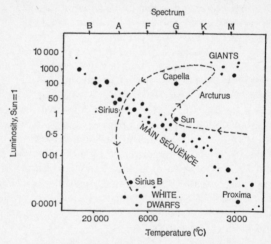

Fig. 17. *Typical H-R diagram. Luminosity is plotted against surface temperature. The giant branch and the Main Sequence are clearly shown. The broken line represents what is now thought to be the evolutionary sequence of a star such as the Sun.*

The theory sounded very plausible, and it was well received, but it too came to grief on account of the time-scale. Ironically, the trouble was the exact opposite of that which had helped to destroy Helmholtz's contraction idea. Instead of being too short, the scale was very much too long. The annihilation process as postulated by Russell would serve to maintain a star not for a few millions or thousands of millions of years, but for something like 10 million million years, which seemed to be quite unreasonable.

The *coup de grâce* was finally delivered by new knowledge about the structure of matter. When only protons and electrons had been recognized, everything was delightfully straight-forward, but with the detection of extra fundamental particles the whole picture became very much more complex. Moreover, it became impossible to picture a particle as a small, solid lump charged with electricity. Quantum theory came to the fore, and an electron took on the dual nature of wave and particle combined. There could be no chance that an electron and a proton could meet head-on and cancel each other out, as Russell had originally supposed, so that the annihilation theory of stellar energy had to be given up.

Such was the situation during the uneasy peace between the two world wars. Contraction was inadequate, annihilation over-generous, and in any case neither process could be regarded as at all satisfactory. It was necessary to find some sort of mechanism involving nuclear reactions, so providing a time-scale of the order of a few thousands of millions of years for stars such as the Sun. The problem was solved, just before the outbreak of the Second World War, almost simultaneously by H. Bethe in America and Carl von Weizsäcker in Germany.

It was found that the essential clue was the behaviour of hydrogen, the lightest and most abundant of the elements. Four hydrogen nuclei can combine to make up one helium nucleus, and during the transformation a little energy is set free and a little mass is lost. The process is not straightforward, but by now it is well understood, and there can be little doubt that we have at last discovered why a star shines. A normal star is using hydrogen as its fuel, and it will continue as a stable luminous body until its supply of available hydrogen starts to run low.

The time-scale for the Sun fell neatly into place as soon as the new theory became accepted, and the age was set as something of the order of 5000 to 6000 million years. Stars which are of initially greater mass squander their energy reserves at a much faster rate, and do not last for nearly so long; S Doradûs in the Large Magellanic Cloud, which has 1 000 000 Sun-power, can hardly maintain its present output for more than a million years to come, which by cosmical standards is a short period indeed. On the other hand, the dim red dwarfs are much more gentle, and have longer life-spans. The old Lockyer sequence of red giant to Main Sequence, Main Sequence to dead globe has had to be given up completely. The only common factor is that in both old and new theories a star begins its career inside one of the gas-to-dust patches known as nebulae. A distinction must be drawn here not only between Herschel's 'irresolvable' and 'resolvable' objects (in other words, gaseous nebulae and external galaxies) but also between the different types of nebulae contained in our own Galaxy. The planetary nebula in Draco, studied spectroscopically by Huggins, is really a faint star surrounded by an immense shell of tenuous gas, and is quite unlike a nebula such as Messier 42 in the Sword of Orion. For stellar birthplaces we must look to the extensive, diffuse gas patches. There are plenty of them; some are spectacular when seen through even small telescopes, and superficially they are far more imposing than the galaxies.

The life-history of a star depends mainly upon its initial mass, and it is convenient to take the Sun's mass as a standard. If the star concerned is of less than 0.1 solar

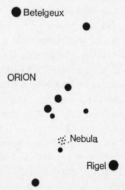

Fig. 18. *Position of the Orion Nebula, which lies below the stars of the Belt and is visible with the naked eye. Betelgeux is a typical red giant, while Rigel is a highly-luminous white star of spectral type B.*

masses, it will condense out of the nebular material, but will never become hot enough at its core to "trigger off" nuclear reactions, so that it will simply fade away until it has lost all its energy.

If, however, the mass is between 0.1 and 1.4 solar masses, things are very different. The "protostar" will be large, diffuse and red; it contracts, with a rise in core temperature, and when a value of approximately 10,000,000 degrees has been reached the star begins to shine because of nuclear reactions taking place inside it. Hydrogen is converted into helium (admittedly in a rather roundabout sort of way), and the star joins the Main Sequence, where it remains for most of its main career. When the supply of available hydrogen runs low, the star has to rearrange itself; different reactions begin, with heavier elements being built up, and the core temperature rises to fantastic values, though the outer surface will expand and cool. The star has become a Red Giant, to the upper right in the H-R Diagram. Eventually, when all nuclear reserves have been used up, the star collapses into a small, very dense White Dwarf, with all its atoms broken and crammed together. A White Dwarf has been described as a stellar bankrupt. Its radiation is very feeble, though the surface temperature is still high. The most famous White Dwarf, the Companion of Sirius, was first observed more than a century ago, though its remarkable nature was not recognized until spectroscopic studies were made from Mount Wilson during the First World War.*

* The first view of the companion was obtained by Clark, in 1862, but the discovery was not totally unexpected. The existence of a faint companion to Sirius had been predicted much earlier by Bessel, who had found that the proper motion of Sirius itself against the background of more remote stars is not uniform; there is a slight 'wobble', and Bessel attributed this, quite correctly, to the pull of an unknown binary companion. Moreover, Clark discovered the companion almost precisely where Bessel had expected it to be. There is some analogy with the case of the planet Neptune in our own Solar System, which was tracked down in 1846 because of its influence upon Uranus. These mathematical triumphs could not have been achieved without following the principles of Kepler and Newton, but by the 19th century the validity of these principles was in any case beyond dispute.

Its density is about 60,000 times that of water, but its diameter is less than three times that of the Earth. Other White Dwarfs are known with even smaller diameters and higher densities.

A solar-type star, then, will become first a giant and then a White Dwarf, until at last all its energy has been used up and it becomes a dead globe. But a star more than 1.4 times as massive as the Sun will have a much more spectacular end. It may well explode as a supernova, blowing much of its material away into space, and leaving a remnant made up of neutrons — even the positively-charged protons and the negatively-charged electrons have been fused together, so that their electrical charges cancel each other out.

Neutron stars are remarkable by any standards. The density may be of the order of 100 million million times that of water, and the old star is spinning rapidly round, producing the rapidly-varying radio waves which have led to the name of "pulsars" for neutron stars in general. It is now thought that a star of even greater initial mass may collapse into something even more extreme than a pulsar, in which case it will be surrounded by a kind of "forbidden zone" from which not even light can escape: this is the concept of a Black Hole.

Very massive stars run through their careers at a greatly accelerated pace, and by stellar standards our Sun is comparatively mild, but the new theories naturally changed all ideas about its future.

This change in outlook naturally affected all theories about the future career of the Sun. For many years it had been recognized that the Sun cannot last for ever. If it had drawn energy from gravitational shrinkage, as Helmholtz had thought, its expectation of 'life' would have been limited to a few tens of millions of years, and all living things on Earth would have been slowly frozen to death as the Sun cooled down and the temperature dropped. On the annihilation theory, the same sort of situation would have prevailed eventually, but it would have been less imminent, and would have been delayed for several millions of millions of years. With the new theory involving nuclear reactions, the Sun's expected life-span is intermediate between these two extremes, but the final end of the Earth will come in a different fashion. Instead of cooling down as it ages, the Sun will join the giant branch, so that it will go through a period of relatively high luminosity before it collapses into a white dwarf. Before then, too, it may have become much less stable than it is now, and the chances of its becoming a variable star cannot be discounted. Any of these events would mean the end of life on Earth, and we can see at once how much our outlook has changed since the 16th century. Then, the Earth was regarded as the supreme body, and the Sun was tacitly assumed to have been created expressly for the benefit of mankind. Now, we know that we are entirely dependent upon the Sun; the idea of our being 'lords of the universe' has become absurd.

Luckily our Sun is a stable star in mid-career, and it is not likely to change much for the next 5000 million years or so. There may well have been minor fluctuations in the past giving rise to the Ice Ages which have affected the Earth now and then throughout geological history, but small variations of this kind are not connected with any long-term evolutionary changes in solar output, and so far as we are concerned the Sun is constant. Eventually, of course, the crisis will come, but it lies so far ahead that speculation is pointless. All we can really say is that according to modern evidence, life on Earth will finally be destroyed by increasing heat rather than by creeping cold.

8. Beyond the Milky Way

By the beginning of the 19th century it had become certain that the Earth is an unimportant planet moving round an equally unimportant star. Herschel, the leading observer of the period, admittedly thought that the Sun lay at or very near the centre of the Galaxy, but he did not for one moment suggest that this could be due to anything but coincidence. The idea of our Earth having been placed in the centre of the universe by some divine agency could no longer be taken seriously.

All through his career Herschel was hampered by his ignorance of the distances of the stars. He did his best to make accurate measures, and it was not his fault that he failed, but the difficulty remained. Since no precise figures were obtainable, he was forced to make some assumptions he knew to be dubious. In particular, he assumed that the brighter stars were relatively near, and that there was a real correlation between distance and apparent magnitude. There is some truth in this, but it is now known that dwarf stars are much more numerous than giants, so that to take a sample of the bright naked-eye stars produces a very misleading selection. For instance, all the 22 apparently brightest stars in the sky (those of the first magnitude) are more luminous than the Sun – 13 giants, and nine bright Main Sequence stars – whereas of the 22 closest stars, only three (Alpha Centauri, Sirius and Procyon) exceed the Sun in luminosity, and only seven are visible to the naked eye.

Though Herschel's assumption was very imperfect, he did manage to draw up a picture of the shape of the Galaxy which was vastly better than anything which had been done before. And as we have noted, he made the bold suggestion that the starry or resolvable nebulae, such as Messier 31 in

Andromeda, might well be independent galaxies, far beyond our Milky Way. This was a brilliant guess, but it was hardly more, and at the time it was incapable of proof or denial. Even the first good measures of star distances, made by Bessel and others following 1838, did not help much. Whatever the starry nebulae might be, they were certainly very distant, and much too remote to reveal any measurable parallax.

The next step was taken by one of the most remarkable astronomers in history – the third Earl of Rosse, an Irish nobleman whose seat was at Birr Castle in County Offaly. The story of Birr is well worth retelling here, because it drives home the striking difference between the situation in the mid-19th century and the position today.

Rosse was not a professional scientist, and had no ambition to become one, but he had great natural gifts, together with patience and enthusiasm. Fairly early in his life he made a 36-inch reflector, and its success encouraged him to go further. In 1840 he decided to build a telescope which would be much larger than anything previously attempted, and he planned a mirror with a diameter of 72 inches, which far surpassed Herschel's 49-inch.

Birr lies in thinly-populated country, and Rosse had no trained helpers, so that he had to do everything himself. First, the mirror had to be cast, which meant building a forge; the days of glass mirrors still lay ahead. After several failures a good 72-inch disk was produced,* and next came the grinding process. Rosse trained labourers from his estate, and turned them into very passable optical mechanics, though naturally he had to do the most delicate operations on his own. The mounting that he devised was clumsy and awkward by modern standards, and the available area of sky was limited to a short distance on either side of the meridian, but within its restrictions it worked satisfactorily, and by 1845 the great telescope was ready. Owing to the unsettled

---

* The Earl's wife, who must have been a most capable person, used the forge for different purposes when it was not being used in telescope-making. In the intervals, the Countess produced excellent iron gates, two of which are still to be seen at the Castle.

state of affairs in Ireland, which took up all Rosse's attention, observing did not begin until two years later, but almost at once there came a spectacular discovery.

Rosse looked at the 'starry nebulae' and found, to his surprise, that some of them were spiral in shape, so that they resembled shining Catherine-wheels. For instance, there was the object listed by Messier as No. 51 in his catalogue, where the spiral form was obvious at a glance – leading to the modern nickname of 'the Whirlpool'. In other cases, such as that of Messier 31 in Andromeda, the tilt was less favourable and the spiral effects were partly masked, but they were still detectable.

Stellar photography had only just begun, and there was no hope of being able to obtain pictures of the spirals. Not only were the photographic processes inadequate, but long time-exposures would have been needed, and the 72-inch reflector was not then mechanically guided (though driving mechanism was added later by the fourth Earl, who was something of an engineering genius). The only course was to draw the spirals, and this was what Rosse proceeded to do. His skilfully executed sketches attracted wide attention, and the term 'spiral nebula' became part of astronomical language.

The interesting point here is that quite recently, not much more than a century ago, it was still possible for an amateur to build the world's greatest telescope and use it to make discoveries of fundamental importance. Actually, the 72-inch remained the largest of all reflectors until the completion of the Mount Wilson 100-inch during the First World War, though it is true that its metal mirror was not to be compared with the glass mirrors made from about 1900 onward, and that the Birr reflector was not extensively used after the 1880s.

Rosse realized the significance of the spiral structure, though at that time no proper interpretation could be made. At first he thought that all nebulae could be resolved into stars, but it soon became clear that there really were two different types – the resolvable and the irresolvable – which presented theorists with a fascinating problem. For years after the initial discovery, there was only one place in the world from which astronomers could look at the spirals: from the grounds of Birr Castle. Rosse's achievement was extraordinary, and nothing of the kind can ever happen again, so that the story of Birr remains unique. Nowadays, the 72-inch mirror is displayed at the Science Museum in London, while the tube, together with the massive stone walls of the 'observatory', can still be seen in the Castle grounds.

Huggins' spectroscopic work, described in Chapter 6, finally confirmed that some nebulae are made up of gas while others are starry. On the other hand, this certainly did not prove that the starry nebulae are external systems, and by 1900 Herschel's speculation had been discounted by most astronomers. It was generally thought that the nebulae, resolvable or irresolvable, were integral parts of our own system, and that our Galaxy must be the only one in the observable universe.

This assumption was perfectly reasonable, and at the turn of the century there was no method of measuring the distances of the spirals, so that astronomers could do little more than speculate. It was during the 25 years following 1900 that the whole outlook changed so dramatically. There has been only one major modification since, and there may never be another.

Everything depended upon improvements in observational techniques. Large refracting telescopes were built, and in 1908 the untiring energy of a young American astronomer, George Ellery Hale, led to the construction of a 60-inch reflector, set up at Mount Wilson in California. This was followed by the famous 100-inch, also at Mount Wilson, which came into operation in 1917, and remained in a class of its own until the completion of the Palomar 200-inch in 1948. The difference between the Rosse telescope and the Mount Wilson reflectors was that the Birr mirror was of metal and the mounting cumbersome, while the new instruments had mirrors of amazing precision together with driving mechanisms so accurate that time-exposures lasting for many hours were perfectly practicable. It is probably true to say that with a large modern telescope the purely mechanical difficulties are as great as those involved in the making and adjustment of the optics.

Equally important were the advances in photography and spectroscopy, without which astronomers would have been hopelessly handicapped. As soon as it became possible to examine detailed photographs of star fields, clusters and nebulae, the scope of investigation was enormously widened. And it was while studying photographs of the Clouds of Magellan, in 1912, that Miss Henrietta Leavitt, of Harvard, made a discovery which was to have far-reaching consequences.

To the everlasting regret of European observers, the two Clouds of Magellan are so far south in the sky that they cannot be seen from any part of the European land-mass — or, for that matter, from California — so that they have to be studied with telescopes set up in more southerly latitudes. They look rather like detached portions of the Milky Way, and both are clearly visible to the naked eye. They contain objects of all kinds, including star clusters and gaseous nebulae, so that they are of special interest from an astrophysical point of view.

Variable stars are also to be found, and there are many of the short-period stars known as Cepheids. Miss Leavitt concentrated upon the Cepheids in the Small Cloud, and found that there was a strange correlation between period and magnitude; the brighter variables always had the longer periods, and there were no exceptions.

At that time the distances of the Clouds were unknown, but there could be no doubt that they were very remote, and this meant that to all intents and purposes it could be assumed that all the objects in them could be regarded as equally dis-tant from us – just as it is approximately correct to say that London and Brighton are the same distance from New York. If, then, the longer-period Cepheids looked the brighter, it followed that they must be genuinely more luminous; and if this applied to the Cepheids in the Small Cloud, it presumably applied to Cepheids everywhere. Miss Leavitt's 'period-luminosity law' was followed up energetically, because it led on to a means of working out the distance of a Cepheid simply by observing its changes in magnitude over a period of a few days or weeks. Once the real luminosity had been found, a comparison with the apparent brightness would give a reliable distance determination.

Many difficulties had to be faced. No Cepheid in our Galaxy is close enough to show a useful parallax, which meant that indirect means had to be employed in working out a standard; moreover, there is the question of the absorption of light by interstellar matter, which was badly underestimated in those early days. But before long the Cepheids came to be regarded as reliable markers, and this is till true today. The period-luminosity law applies only to the Cepheids, while other short-period variables known as RR Lyrae stars have a law of their own (Fig. 19).

Harlow Shapley, also of Harvard, studied the variables in the remarkable objects known as globular clusters, which lie around the main part of the Galaxy, and of which rather more than 100 are known. The globulars are huge systems, but they are so remote that only a few are visible with the naked eye; the three brightest are Messier 13 Herculis in the northern hemisphere, and Omega Centauri and 47 Tucanæ in the far south.

It had long been known that the globulars are not uniformly distributed in the sky, and that there is a marked southern concentration, notably towards the region of Sagittarius. By studying the variable stars in the globulars, Shapley was able to fix the distances of the clusters themselves, and hence to work out the dimensions of the Galaxy. He found that the overall diameter was likely to be about 300 000 light-years, and that the Sun lay well away from the centre, thus giving us an asymmetrical view of the system of globulars. Very understandably, Shapley had underestimated the amount of interstellar absorption, and the modern value for the diameter of the Galaxy is 100 000 light-years, with the Sun lying some 32,000 light-years from the centre, but this is merely an adjustment; the basic conclusions have been found to be correct.

However, Shapley did not then regard the spiral nebulae as

extragalactic, and there was one argument which seemed highly plausible even though it is now known to be wrong. In August 1885 a new star had been observed in the Andromeda Spiral; it had reached the sixth magnitude, so that for a while it was just visible with the naked eye.* Of course, there was no

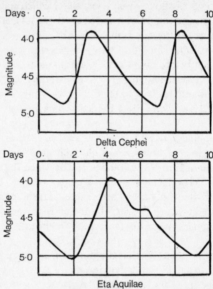

Fig. 19. *Light curves of the two Cepheid variables. Delta Cephei has a period of 5.3 days. Eta Aquilae just over 7 days. From Earth, the two stars appear almost exactly of the same brilliancy, but Eta Aquilae has the longer period; it is therefore the more luminous of the two, and presumably more remote (though various complications, notably the absorptions of light in space, have to be taken into account).*

proof that the nova belonged to the Spiral; it might have simply flared up in the same line of sight, and it is worth noting a comment by G. F. Chambers, a well-known astronomical writer of the period, to the effect that when he looked at the nova he concluded that 'the star had nothing to do

* It was discovered by an Irish amateur, Isaac Ward, and independently by a Hungarian baroness who had taken a casual look at the Spiral through a small telescope belonging to a guest in her house.

with the nebula'. However, the general consensus of opinion
was that the nova really did lie in the Spiral, and others had
been detected in later years, though none had become so
bright. If the Spiral were comparable in size with our Galaxy
it would have to be about 10 000 000 light-years away (remember that at that time, our Galaxy was thought to be
three times the size that it really is). Yet if novae in the Spiral
were similar to our own novae, they could not be anything
like 10 000 000 light-years away – unless they were much
brighter than the novae seen in our Galaxy, which did not
appear to be very logical.

The 1885 star is now known to have been a supernova, very
much more luminous than an ordinary nova, so that it was an
exceptional object. The other novae seen in spirals were of the
usual variety, and the discrepancy was due to the over-
estimation of the size of the Milky Way system.

The whole question of the status of the starry nebulae
remained unsolved until 1924, when E .E. Hubble tackled the
problem with the help of the 100-inch Mount Wilson reflector.
He searched the spirals for short-period variables, both
Cepheids and RR Lyrae stars, and at last he managed to
locate Cepheids in the Andromeda Spiral. At once it became
clear that the distances were truly immense. Hubble gave the
distance of the Andromeda system as 900 000 light-years,
though before long this estimate was reduced to 750 000
light-years.

In any case, the Spiral lay well beyond the edge of the Milky
Way. It and others of its kind could only be external systems,
so that our Galaxy proved to be one of many. First the Earth,
then the Sun had been shown to be unimportant in the universe
as a whole; now the same treatment was meted out to the
Galaxy. Not surprisingly, the evidence was accepted at once.
There were no attempts to cast doubt upon Hubble's observations, or the conclusions to be drawn from them. The facts
were plain, and could not have been disputed even if anyone
had wished to do so. Human thought had made great progress
in the three centuries since Galileo's time.

Yet even when everything had been taken into account, an

annoying discrepancy remained. If the Andromeda Spiral lay at 750 000 light-years, as indicated by its Cepheids, its globular clusters were significantly smaller than those of our Galaxy, and its novae somewhat fainter. Moreover, no RR Lyrae stars had been found there. The RR Lyrae variables – formerly known as cluster-Cepheids, though by no means all of them are cluster members – are not so luminous as the classical Cepheids, but they are around 100 times as powerful as the Sun, so that they might have been expected to show up in the Andromeda Spiral. To suppose that they might be absent altogether did not seem reasonable. Finally, and most significant of all, the distance measures indicated that our own Galaxy must be considerably larger than any other.

Not very many years earlier, the idea that we live in a particularly large galaxy would have seemed only right and proper, but by the inter-war period astronomers had become wary of coincidences of this sort. Then, in 1942, the late Walter Baade undertook a new study of the Andromeda Spiral, again using the Mount Wilson 100-inch, but with the added advantage that city lights were blacked out as a war-time measure, thereby making the sky appreciably darker than usual and allowing longer time-exposures to be made at the telescope without the risk of the plates becoming fogged by scattered light. For the first time it became possible to resolve not only the arms of the Spiral, but also the centre. Baade found that in the nucleus the brightest stars were red giants, while in the arms the leaders were hot bluish stars of the Main Sequence. When he published his results he introduced a new term in astronomy, 'stellar population', Population I being made up of stars in the arms and Population II consisting of the stars near the nucleus.

The significance of this discovery became apparent after 1948, when the 200-inch reflector at Palomar came into use. Other galaxies, too, proved to be made up of these two Populations, but there were some galaxies which seemed to be entirely Population II, and this applied also to the globular clusters. It was found that the vital rôle was played by inter-stellar material. Population I areas were rich in gas and dust, indicating that star formation was still going on; in Population II regions there was relatively little gas and dust, so that star formation had evidently ceased. This was very satisfactory, because red giants are more advanced in their careers than Main Sequence stars. If Population II were made up of old objects, it would be logical to expect its brightest members to be red giants which had already left the Main Sequence.

Baade's suspicions were aroused, and he began to compare the Population I Cepheids in our own part of the Galaxy with the Population II Cepheids in the globular clusters. Again he found a discrepancy, and he realized that the Population I Cepheids were the more luminous of the two classes. This at once gave him a clue to the error in the distance-measure of the Andromeda Spiral.

Miss Leavitt's period-luminosity law had been worked out on the assumption that all Cepheids were of the same type, and this was now found to be incorrect. Shapley's estimates for the size of the Galaxy were based on the original law, and had not been affected − but the distances of the Cepheids in the arms of the Andromeda Spiral had been calculated on the assumption that they were of Population II, whereas in reality they were of Population I.

It followed that since these Cepheids were more luminous than had been thought, they must also be more remote. Almost overnight, the calculated distance of the Andromeda Spiral had to be doubled, and the same applied to all the other galaxies, so that the observable universe was twice as large as had been believed. Further adjustments have been made since, and the modern value for the distance of the Andromeda Spiral is 2 200 000 light-years. This removes all the discrepancies; the globulars, the novae, the supergiants and all the other objects in the Spiral become strictly comparable with those in our own system. Everything fits in so neatly that we may be confident that there will be no further fundamental change in our estimates.

In this respect it seems therefore that the final step has been taken. Our Galaxy is completely normal, and is not of exceptional size. Indeed, the Andromeda Spiral is 1½ times larger in diameter, and contains more than our own quota of 100 000 million suns. We can also understand why the RR Lyrae stars did not show themselves there: because they are not so powerful as the Cepheids, they were lost in the general background glow of the Spiral.

It would be quite wrong to suppose that all the external galaxies are spiral in shape. Some are elliptical, some are globular – so that superficially they look not unlike globular clusters – and some are irregular; such are the two Clouds of Magellan, which are about 180 000 light-years away, and may be regarded as junior companions of our Galaxy. (Indications of spiral structure have been claimed, but are certainly not well-marked.) Even the spirals are of different kinds, ranging from the 'loosely-wound' systems to galaxies with large nuclei and tight, closely-knit arms. It has also been proved that our Galaxy is spiral, but since the main proofs were provided by radio-astronomy methods they are best discussed in Chapter 9.

Much has been learned about the evolution of the stars, and there can be little doubt that the basic ideas are correct; for instance, it is now inconceivable to suppose that a white dwarf may be younger than a red giant, or that a red giant will evolve into a Main Sequence star. Unfortunately, knowledge about the evolution of a typical galaxy is very slight. It is tempting to suppose that a spiral may evolve into an elliptical, or vice versa, but there are serious objections; in particular, the giant ellipticals are far more massive than the largest of the spirals. We do not even know why spiral arms form, or whether they are permanent features. To make any attempt to discuss matters of this sort would be out of place in a book dealing only with the development of astronomical thought, and it must suffice to repeat that between 1924 and 1952 the fundamental problem of the status of our Galaxy was solved.

Powerful though they are, the Cepheids can be observed out only to a relatively few millions of light-years. The highly-luminous supergiants can be detected out to greater distances, and estimates can be made on the reasonable assumption that the brightest supergiants in our Galaxy are about equal to the brightest supergiants in other galaxies. Further away still, we have recourse to estimating distances by assuming that galaxies of similar shape are also likely to be of similar size. The results of these indirect methods are bound to be subject to error, but they are far better than nothing at all.

Then, too, there is the spectroscopic method based on the red shift. Here we are starting to delve into the problems of cosmology, but it may be helpful to give a brief introduction here.

Even before Hubble's work had shown the galaxies to be external, spectroscopic studies had been started by V. M. Slipher and his colleagues at the Lowell Observatory in Arizona. The spectra of the galaxies were clear enough to be measured, and it was found that with a few exceptions all the shifts were to the red. Interpreted as a Doppler effect, this meant that the galaxies were receding from us.

The exceptions to the general rule were found in the very nearest of the galaxies, notably the Clouds of Magellan, the Andromeda Spiral and the fainter spiral in Triangulum. These, with other much smaller systems discovered later, are now known to make up what is termed the Local Group. But for the vast majority of the galaxies, the red shifts were extremely pronounced. After Hubble's revelations of 1924 the situation became even more interesting, because it appeared that the velocity of recession increased with increasing distance. The more remote galaxies were racing away at the greater speeds, so that the whole universe seemed to be in a state of expansion.

Together with his colleague, M. Humason, Hubble drew up a definite relationship linking distance with recessional velocity. Assuming that this held good throughout the observable universe, it provided an excellent way of estimating the distances of galaxies which are so remote that no individual stars can be made out – and the method is still in use today. If, for instance, the red shift of a galaxy shows it to be receding at 26 000 miles per second, Hubble's Law indicates that its distance must be 700 000 000 light-years.

There are obvious dangers in this procedure, because we depend upon two assumptions: that the red shifts are pure Doppler effects, and that Hubble's Law remains valid far out across the universe. More will be said about this below. However, the general consensus of opinion at the present time is that the expanding universe theory is correct.

Just as our Galaxy is a member of a definite group, so we know of many other groups or clusters of galaxies, some of which contain hundreds of systems. Individual groups do not expand, and may be regarded as comparatively stable units, but each group is receding from each other group, and there is no suggestion that our Galaxy lies at the centre of the universe. The expansion is 'universal' in every sense of the word. The time when we can consider ourselves as occupying any special position has long since passed.

9. The Visible and the Invisible

Until less than 40 years ago, astronomy was purely optical. All our knowledge was gained by means of the light sent to us from the various bodies in the sky, and everything depended upon the telescope, which had to collect the light for analysis. Quite obviously this imposed severe restrictions, because visible light makes up only a very small part of the electromagnetic spectrum. The visible range extends from about 3900 Ångströms for violet up to about 7500 Å for red (1 Ångström unit is equal to a hundred-millionth part of a centimetre). And since stars and other objects emit radiation in the visible range, there was no reason to doubt that they emitted in other parts of the electromagnetic spectrum as well.

It is worth noting that it had been an astronomer, Herschel, who had been responsible for proving the existence of infrared radiation with wavelengths too long to affect our eyes. He had examined the heating powers of various parts of the spectrum, simply by passing light through a prism and then using a thermometer in different parts of the rainbow band. He had found perceptible effects well beyond the red end of the visible spectrum, and he had realised that he was dealing with emissions of precisely the same kind as those which we can see. Yet at that time there was no thought of infra-red astronomy, and still less of the astronomy of radio waves.

The first real inkling of what we may term 'invisible astronomy' came in 1931, when a young physicist named Karl Jansky, working for the Bell Telephone Laboratories in America, was carrying out research into problems of the static which causes interference with radio transmission. For this purpose he had set up an experimental aerial of unusual design, usually nicknamed the 'Merry-go-Round'. During his

investigations he picked up a strange hissing noise which at
first defied identification, but which he subsequently found to
come from the Sagittarius region of the Milky Way.

This turned out to be one of the most momentous discoveries
in the history of astronomy, because it led on to a new field of
research which has now become of the utmost importance. Yet
Jansky was curiously unexcited. He published a few papers,
but most of them came out in radio and electronics journals,
so that astronomers in general did not come across them.
Moreover, Jansky soon transferred his attention to other
work, and never followed up his discovery as he might have
been expected to do. It was only in 1937 that an American
amateur, Grote Reber, made the first intentional radio telescope – a parabolic metal mirror $31\frac{1}{2}$ ft in diameter – and began
a systematic survey of the sky to see what radio sources he
could detect.

He found several, but they were not coincident with bright
stars, and seemed to come from regions of the sky unmarked
by any conspicuous objects. Like Jansky, he published some
papers. Also like Jansky, he was more or less ignored. Then
came the war, and the discovery by a British team headed by
J. S. Hey that some peculiar 'jamming' of radar equipment
came from the Sun and not, as had been suspected, from
Germany. The Sun, too, was a source of radio emissions.

The story of early research in radio astronomy is certainly
strange, and probably without parallel. A fundamental
discovery had been made in 1931; the techniques of the period
were quite adequate for it to be capitalised – and yet 14 years
later virtually nothing had been done, except by a single
amateur and, unintentionally, by a military research unit.
Astrophysicists and cosmologists had been hard at work
observing and theorising, but they had paid no attention
whatsoever to the powerful new weapon which they could
have used. The underlying reason may well have been that the
idea of picking up invisible radiations from space was entirely
novel, and at least some astronomers may have been inclined
to dismiss it in the same way as space research was regarded
with distrust in later years. Whether this is true or not, there

can be no doubt that progress was extraordinarily slow. It was only after the end of the war that radio astronomy became an accepted and 'respectable' study.

Despite this tardy beginning, there were striking developments in the immediate post-war years. Various radio telescopes were built, some of them of the Reber paraboloid or dish type, others quite different in aspect. By 1951 a hundred separate radio sources had been found in the sky, and a few of them had been identified with visible objects. Initially, the main trouble was that the resolving power of a radio telescope is bound to be low in comparison with that of an optical telescope, so that there was great difficulty in finding accurate positions for the sources, and two sources lying side by side were recorded as only one. What may be regarded as the first major triumph was the detection of emission at a wavelength of 21·1 centimetres, by Ewen and Purcell in America; it proved to come from the clouds of hydrogen spread through the Galaxy, and studies of the motions and distribution of these clouds gave the final confirmation that the Galaxy is truly spiral in shape. The 21-cm radiation had been predicted during the wartime years by H. van de Hulst in Occupied Holland, but confirmation was inevitably delayed.

Originally the discrete sources were known as radio stars, but the term was obviously inappropriate, simply because the sources were not stars in the usual meaning of the word. Some of the sources lay inside our Galaxy, but others appeared to be external. The only individual star known to be a radio source was the Sun, whose emissions were detectable only because of the comparatively small distance involved.

The Crab Nebula was found to be a powerful source. As was already known, this remarkable object is the wreck of the supernova observed in the year 1054 (Fig. 20), though since it is some 6000 light-years away the actual outburst occurred long before. Another strong source, in Cassiopeia, was associated with faint gas-clouds which also seemed likely to be supernova debris. On the other hand, a source in Cygnus was identified with a peculiar object which was undoubtedly external, and was taken to be a galaxy of unusual type. As radio equipment was improved, more and more sources were identified with visible objects, until nowadays the list is very long indeed.

Just as a large optical telescope is needed to show faint and remote objects, so a large radio telescope is essential for the study of weak sources – and, moreover, the larger the radio telescope the greater the resolving power. In 1955 the great 250-foot paraboloid at Jodrell Bank in Cheshire was completed; its success was due to the skill and confidence of Professor (now Sir) Bernard Lovell, whose rôle in radio astronomy has been very similar to that of Hale in optical

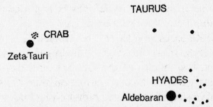

Fig. 20. *Position of the Crab Nebula, near the star Zeta Tauri in the constellation of Taurus (the Bull). The Crab is the remnant of the supernova seen in the year 1054, which for a time became brilliant enough to be seen in broad daylight.*

astronomy. For many years the Jodrell Bank "dish" was in a class of its own.

To give even a brief outline of the development of radio astronomy over the past few decades would be beyond the scope of the present book, and all that can be done is to mention the radio studies in relation to cosmological thought. Obviously, the first task was to decide just what the radio souces could be. Some lay within our Galaxy, and of these many turned out to be supernova remnants; the Crab Nebula was one. Then, in 1967, strange, rapidly-varying radio sources were discovered, and were named pulsars; we have now found that these are associated with old supernovae, and there is one in the Crab Nebula. The first pulsar was found by Miss Jocelyn Bell, at Cambridge, and for a time it presented astronomers with a true puzzle! Today several hundreds of pulsars are known, though only two have been identified with optical objects. One of these is, of course, the Crab pulsar. The other is in the southern constellation of Vela. The Vela pulsar, contained in the so-called Gum Nebula (named in honour of the late Dr. C.S. Gum) is actually the faintest optical object ever recorded; its magnitude is below +24. It was first seen in 1977 by a team working with the new Anglo-Australian reflector at the Siding Spring Observatory in New South Wales.

There are also other galactic radio sources, including a few individual stars. However, it is the extragalactic sources which are of such fundamental significance in cosmology.

However, it is the extragalactic sources which are of such fundamental significance in cosmology.

Around 1960 it was thought that sources of the Cygnus A type were galaxies in collision. On this theory, two galaxies belonging to the same group were passing through each other, moving in opposite directions; the individual stars would seldom or never collide, but the constant interaction between the interstellar particles would produce the radio emission observed. Unfortunately it was then found that any such process would be hopelessly inadequate, and the whole idea of colliding galaxies had to be abandoned. So far no really plausible alternative has been put forward, and neither can we account for the fact that certain galaxies are strangely powerful at radio wavelengths while others are not. Some radio galaxies, such as Messier 82 in Ursa Major, show evidence of great past explosions within their nuclei, and this is undoubtedly significant, though once again the causes are completely unknown.

Our worst handicap is that we are still ignorant about the evolutionary course of a typical galaxy. Whether an elliptical system can ever turn into a spiral, or vice versa, we do not known – though most modern astronomers doubt it. Also, does every large galaxy pass through what we may term a 'radio stage'? Here too we can give no definite answer.

By 1963 the existence of radio galaxies had been established beyond all doubt, and it seemed that the only major problem was to account for the strong emissions. There was apparently nothing to distinguish a radio galaxy from a 'radio-quiet' galaxy apart from its energy in the long wavelengths, and there was certainly no thought among astronomers that objects of an entirely new class might exist. Stars, planets, galaxies, clusters of galaxies – all these were to be found in the universe, and anything of fundamentally different type was regarded as improbable in the highest degree. Then came the discovery of the quasars, which, if the present majority view is accepted, are the most luminous and the most puzzling objects known.

Certain radio sources seemed to coincide in position not with supernova remnants or external galaxies, but with what looked like faint blue stars. One of these sources, 3C–273 (the 273rd object in the third Cambridge catalogue) was so placed that it could be occulted by the Moon, and therefore its position could be measured very accurately. It proved to be a double source, with a separation of 20 seconds of arc. Armed with this information, Maarten Schmidt at Palomar examined the spectrum of the 'star' concerned, and found, to his surprise, that the spectrum was unfamiliar. Not only was it quite unlike that of a normal star, but there was a very large red shift, indicating a high rate of recession and so, presumably, a great distance. This was the first quasar; others were soon identified.

If quasars are as remote as their red shifts indicate, they must be much more powerful than ordinary galaxies. Yet they appear small, and in some cases almost stellar. (One quasar has been identified on photographs taken almost a century ago, so that it cannot be said to have been unknown, though of course, it had always been mistaken for a star.) The immediate problem is to explain how a relatively small body, only a few light-years in diameter, can radiate with such stupendous energy. There is a growing body of opinion that a quasar may be associated with the nucleus of a very active galaxy, and we have also to reckon with the so-called BL Lacertae objects, which are probably linked with quasars and which have almost featureless optical spectra. At the moment we have to admit that our knowledge of the nature of all these very remote, highly luminous objects is extremely incomplete.

It has been claimed, no doubt justifiably, that the identification of quasars has been the most important event since Hubble established that the spiral nebulae lie beyond the Galaxy. Quasars were so unexpected, and so enigmatical. Yet astronomers had no hesitation in accepting the situation; they did not dispute the observational results, and they set to work to explain them. There is a changed attitude here, as is brought out by recalling another 'incredible' puzzle of half a century earlier: that of the white-dwarf companion of Sirius.

Something has already been said about white dwarfs. They are very old stars, whose nuclear reserves have been exhausted, and whose matter is degenerate, so that the fragments of atoms are packed together with almost no waste space. The companion of Sirius was the first to be discovered, and was assumed to be a cool red star of normal size. Its distance was known to be 8·6 light-years, and it was found to send out only 1/10 000 the light of its brilliant primary. Studies of its orbit showed that it is almost as massive as the Sun. If it were hot and white, it would also have to be very small, and it is worth quoting a paragraph from a well-known book called *Astronomical Essays*, written in 1907 by J. E. Gore:

'If its faintness were merely due to its small size, its surface luminosity being equal to that of our Sun, the Sun's diameter should be the square root of 1000, or $31\frac{1}{2}$ times the diameter of the faint star, in order to produce the observed difference of light. But on this hypothesis the Sun would have a volume 31 500 times the volume of the star, and, as the density of a body is inversely proportional to its volume, we should have the density of the Sirian satellite over 44 000 times that of water. This, of course, is entirely out of the question. . . .'

Accordingly, Gore concluded that the companion of Sirius must be relatively large, cool and rarefied. Anything else seemed impossible. When, during the First World War, spectroscopic work showed that the surface temperature was considerably higher than that of the Sun, astronomers were openly sceptical. It took some time for them to 'accept the impossible', and concede that densities of this order did actually occur.

A star with density 44 000 times that of water was discounted in 1907, and yet it was true. This shows that we cannot afford to be sceptical about the quasars simply because they appear incredibly luminous. Shakespeare's 'there are more things in Heaven and earth, Horatio . . .' may have become somewhat hackneyed, but it is extremely apt in cases such as those of white dwarfs and quasars.

Of course, over-credulity is every whit as dangerous as total scepticism, and there have been rational attempts to explain the quasars as comparatively local objects, possibly ejected from the nucleus of our Galaxy by a super-explosion between 10 million and 100 million years ago. However, this does not seem very plausible, because an explosion of such magnitude would spread matter outward in all directions rather than eject small, highly luminous bodies (even if the explosion itself could be explained). The other possible solution which would reduce the immense distances of the quasars is that the red shifts in their spectra may not be Doppler effects. There are serious difficulties here too, but again we are touching upon cosmology, so that further discussion is best postponed until Chapter 10.

Quasars were discovered only because of their radio strength; if they had not been so powerful at long wavelengths, they would still be mistaken for ordinary stars in the halo of our Galaxy. There is no need to stress the importance of the new field of research opened up by Jansky's chance discovery in 1931. But as well as emitting at long wavelengths, bodies in the sky also radiate in the short-wave part of the electro-magnetic spectrum, and here too there have been striking developments during the past few years.

Below the shortest wavelengths visible to the eye we come to the ultra-violet, and then successively to X-rays and the remarkably short, highly penetrating gamma-rays. The main trouble about studying them is that the upper part of the atmosphere acts as an effective shield, and the short wavelengths are prevented from reaching ground level. From our point of view this is just as well, because some of the radiations are harmful, and but for the screening it is likely that advanced life on Earth could never have developed, but to the astronomer the atmosphere imposes severe restrictions. Radiations can reach us only through what is termed the optical window (visible light) and the radio window (certain wavelengths in the radio range). All the rest are blocked out.

This meant that before the development of rocket methods, very little could be done. Instruments could be flown in high-altitude balloons, but the most inconvenient of the layers are beyond the ceiling of any balloon, and something more powerful is needed.

Rockets, of course, are not new. Firework rockets date back for many centuries, and military rockets were used now and then until the latter part of the 19th century, though it cannot be said that they were uniformly successful. (During the Napoleonic Wars, there was a Rocket Mounted Corps of the British Army.) But the simple, solid-fuel scientific rocket has marked limitation, and it was not until the work of the Russian pioneer, K.E. Tsiolkovskii, that any really useful theoretical progress was made.

Tsiolkovskii's first papers, dating from 1903, aroused little interest because so few people read them — they were published in obscure Russian journals, and were not translated until much later. Gradually, however, the idea of liquid-fuelled rockets took hold. In 1919 R.H. Goddard, the American "father of rocketry", published a monograph entitled *A Method or Reaching Extreme Altitudes*, in which he suggested that rockets might well be used to explore the upper atmosphere, and to carry out research which could never be attempted from ground level. He went on to propose that it could be possible to send a charge of flash-powder to the Moon, and it was this part of his paper which was seized upon by the Press. Goddard, a reticent man, was irritated, and made no more public announcements, so that when he fired the first liquid-propellant rocket, in 1926, not many people heard about it.

Meanwhile, interest had been aroused by a book by Hermann Oberth, of Roumania: *The Rocket into Interplanetary Space*. Practical research began in Germany, led by men such as Wernher von Braun, and this led on to the unwelcome development of the V.2, Hitler's terror-weapon which was used to bombard England during the last part of the war. However, scientific work was also undertaken by various rocket teams during the 1930s, when spectrographic equipment was carried aloft and recorded the Sun's spectrum well into the ultra-violet and even into the X-ray region.

After the overthrow of the Third Reich, the scientific rocket began to make its first real impact upon astronomical thought. There was, of course, the concept of sending men to the Moon, which had for so long been ridiculed; but the last sceptics were silenced on 4 October 1957, when the Russians launched the first artificial satellite, Sputnik 1, and ushered in the Space Age. Yet the first major discovery made from an artificial satellite was American. Explorer 1, launched by the team headed by Wernher von Braun, was responsible for the detection of the radiation belts round the Earth now known as the Van Allen Belts in honour of their chief investigator.

Since those early days, the primitive satellites have been succeeded by sophisticated orbiting observatories and even manned space-stations, of which Skylab was probably the most famous (despite its somewhat inglorious end in 1979, when it disintegrated in the upper air and showered fragments over parts of Australia) though the Russian Salyut stations have been of equal value to science. Obviously, a satellite has a tremendous advantage over a rocket which can stay about the shielding atmosphere for only a few minutes before falling back to destruction. Observers in Skylab and the Salyuts have carried out a great deal of pioneer research, concerned mainly with what is called "invisible astronomy" — both long-wave (radio astronomy) and short-wave (ultra-violat, X-ray and gamma-ray research).

Actually, X-ray astronomy began in June 1962, at the White Sands Rocket Base in New Mexico, when a vehicle equipped with X-ray detectors was sent up. The results were of intense interest inasmuch as X-rays were found to be coming from the centre of the Galaxy. In the following year H. Friedman and his colleagues were even more successful, and traced two X-ray sources, one of which lay in Scorpio (Scorpio X-1) and the other coincided with the position of the Crab Nebula. Since then many discrete X-ray sources have been found, and special satellites have been launched to study them. Of these unmanned vehicles, Ariel 5 and Ariel 6, both of British construction (though American-launched) have been particularly notable.

Gamma-rays, even shorter than X-rays, have also been studied by equipment carried in rockets and satellites. The difficulties are very great, and to date (1979) it has been calculated that the number of gamma-ray photons recorded by scientific equipment is no greater than the number of light-ray photons received from a star such as Vega in a single second! Once again the Crab Nebula is of the utmost importance, but gamma-ray sources are certainly of various different types.

It is fair to comment that the period between 1957 and the end of the 1970s was a time of great discoveries. Objects were found which had been not only unknown, but in some cases unsuspected. Quasars have already been mentioned; but for their radio emissions they would never have been identified. In 1967 came the detection of the rapidly-varying radio sources which are now known as pulsars, and which have been found to be neutron stars — the remnants of very massive stars which have collapsed. A typical neutron star may be only a few miles in diameter, with a mean density of 100 million million times that of water; it is spinning round very quickly, and this explains the rapid radio pulses, since the overall picture may be loosely compared with that of a rotating searchlight; every time the Earth passes through the "radio beam", a pulse is received. It is true that the basic concept of a neutron star was suggested as long ago as 1934, by W. Baade and F. Zwicky, but until the development of new techniques such ideas could be no more than theoretical

Then, of course, there are the enigmatical Black Holes. Not all astronomers believe in them; but according to the majority view, a Black Hole is a "forbidden area" surrounding a collapsed star, even denser and smaller than a neutron star, whose escape velocity is so high that not even light can break free. One strong candidate is Cygnus X-1, a system made up of a white supergiant star, HDE 226868, with a mass 30 times that of the Sun, together with an invisible companion body which has 14 times the Sun's mass. It is an X-ray source, and it is thought that the X-radiation comes from material which has been pulled away from the supergiant and is about to be drawn into the Black Hole companion.

Without "invisible astronomy" our knowledge would have remained very fragmentary, and it is very clear that new methods involving rocketry and space research have become all-important. It is wrong to suppose that such methods can ever supersede optical astronomy. They are tackling the same problems from different view-points. Neither must we forget another approach which has a distinctly novel flavour. The Sun should in theory emit large quantities of neutrinos, which are particles with no mass and no charge, and are so penetrating that ordinary equipment cannot record them. Huge tanks of "detecting liquid" — in fact, cleaning fluid! — have been placed deep underground, so that only neutrinos can reach them. As yet the results are baffling, since the Sun's neutrino flux appears to be dramatically less than had been expected; but the work is still in its preliminary stage.

The astronomer of pre-war days was essentially a physicist and a mathematician as well as an observer but today many other branches of science have joined in the overall programme. Astronomy used to be regarded as a subject on its own, so to speak. This was never true, but it is demonstrably false today; a few decades have led to a radical change in our outlook.

10. Cosmologies Old and New

Much has been written lately about the origin of the universe. Some authorities have claimed that the universe is in a steady state, so that it had no beginning and will have no end; others have maintained that there was a definite moment of creation, many thousands of millions of years ago, and that the universe will eventually die. Now that the steady-state idea has been abandoned, there is some support for an oscillating or cyclic universe. Yet it seems that there is a general confusion between 'origin' and 'development'. We know nothing whatsoever about the creation of the universe, and nobody (apart from Biblical Fundamentalists) will claim otherwise. What has been discussed at such length is the development of the universe, which is a very different matter.

One really useful piece of information is the age of the Earth, which we know, fairly accurately, to be between 4500 and 5000 million years. It is logical to assume that the Sun is older, and the Galaxy older still, so that the universe itself cannot have an age of less than 10 000 million years; no doubt it is substantially greater than this, but at least we can be confident about our lower limit.

The first problem to be tackled is that of the origin of the Solar System, and here, for once, ideas have shown something of a drift back to the theories of long ago. In the last years of the 18th century Pierre Simon de Laplace proposed his 'nebular hypothesis', in which the Solar System began as a rotating gas-cloud and evolved slowly into the planetary system of today. Laplace supposed that as the gas-cloud contracted it produced gaseous rings, and that each ring condensed into a planet, with the last remnants of the cloud making up our modern Sun. However, the mathematical objections were found to be so serious that the whole theory was later given up. Textbooks of the period before the last war refer to it as an historical curiosity, but not as an idea to be taken seriously. It is true that the original Laplace picture is untenable, but the theories most favoured today resemble it much more closely than might have been expected.

The next step was taken by Chamberlin and Moulton, in America, who worked out a completely different hypothesis, and published it in 1900. They introduced a wandering star, which was thought to pass relatively close to the Sun, raising massive tides until large quantities of matter had been torn away. After the intruder retreated, the Sun was left surrounded by a cloud of débris; now and then aggregations were formed, and once these aggregations reached a certain critical size they would have enough gravitational pull to draw in extra material. The final result was that most of the débris was collected into the bodies we now know as the planets.

There were various modifications of the original tidal theory. Sir James Jeans, for instance, put forward the idea that the material from the Sun was torn away in the form of a cigar-shaped tongue of matter, and this sounded plausible, since the largest planets (Jupiter and Saturn) would have come from the middle of the 'cigar' (Fig. 21). A. W. Bickerton preferred a grazing impact between the Sun and the wandering star, and this picture was worked out in more detail by Sir Harold Jeffreys. According to H. N. Russell (of Hertzsprung–Russell diagram fame) the Sun used to have a binary companion, and it was this companion-star which was struck by the intruder, planet-forming material being scattered in the process; R. A. Lyttleton considered that the effect of the wandering star was to wrench the binary companion away from the Sun altogether. In fact, by the end of the war there were as many tidal theories as there are planets. Yet the mathematical difficulties remained, and by now all ideas involving a passing star are definitely out of favour.

Today there seems little doubt that the Earth and the other planets were formed from a "solar nebula", a cloud of dust and gas associated with the youthful Sun. Not all authorities agree upon the details, but the general principles are fairly well-established. One relevant point stands out at once. On the old tidal theories, planetary systems would be extremely rare, because individual stars are always widely separated in space, and collisions or near-collisions must be remarkably uncommon; indeed, our Solar System might be the only one in the Galaxy. With modern theories, the situation is quite different. What can happen to the Sun can presumably also happen to other stars, in which case planet families may be expected to be plentiful.

There is strong indirect evidence in support of this. We cannot see planets of other stars, because they are much too faint, but in a few cases we can detect the gravitational effects of relatively massive bodies — presumably planets — associated with comparatively close and lightweight stars. For instance, Barnard's Star, a dim Red Dwarf only 6 light-years away, shows irregularities in its proper motion which seem to be due to either one massive planet, or (more probably) two less massive planetary attendants. All in all, there is no reason to believe that our Solar System is unique, and it may even be that most solar-type stars are accompanied by planets.

Fascinating though it may be, the origin of the planets is only a minor part of a much larger problem. When we come to inquire into the development of the universe, we have even less positive information to use as a basis. The one inescapable fact is that the material making up everything in the universe has real existence, and so it must have been created in some manner or other. It is here that our ignorance is so complete. The only escape-route is to assume that the universe has always existed. Certainly this removes the need for an act of creation, but it introduces an equally serious difficulty, because we have to try to consider a period of time that had no beginning. It is best to admit that so far we have no solution.

Of the classical ideas still in vogue today, the first is that known commonly as the 'big bang', due largely to a French abbe, Lemaitre, and developed and made known by one of Britain's greatest 20th-century astronomers, Sir Arthur Eddington. The material in the universe is supposed to have been created at one specific moment, perhaps 15,000 million years ago, and to have taken the form of a primaeval atom' of amazing density. The primaeval atom exploded, and sent material outward in all directions. After a period of uneasy balance between two opposing forces, gravitation and cosmical repulsion, the expansion continued; galaxies formed out of the material, stars out of the galaxies, and so on. The expansion is still in progress, and will continue indefinitely.

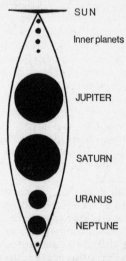

Fig. 21. *Jeans' theory of the origin of the planets. It was thought that the planets were drawn out of the Sun by the action of a passing star. The theory has now been rejected, plausible though it sounded when first put forward.*

Cosmical repulsion was taken to be the exact opposite of gravitation. There is no experimental evidence in favour of it, and its nature was admitted to be unknown, but this in itself was no serious obstacle, because we are equally ignorant of the nature of gravitation even though we are so familiar with gravitational effects. It was assumed that while gravitation weakens with increasing distance between two objects, cosmical repulsion grows stronger, so that in the universe as a whole it is the cosmical repulsion which has the upper hand.

This was the original form of the "big bang" theory. It went through many modifications, but the essential principles were retained: a universe which was created at one moment in time, is now evolving, and will eventually die. Important confirmation was obtained by American researcher in 1965-6, when they identified what may be called "background radiation" coming from all directions, at a temperature of 3 degrees above absolute zero; presumably this represented the remnant of the original "big bang".

Meantime there had been the rise and fall of a completely different theory, proposed by H. Bondi and his Cambridge colleagues soon after the war. This assumed that the universe must be in a steady state, and that it had an infinite past and an infinite future, so that old stars and galaxies were replaced by new ones formed from material created spontaneously out of nothingness. The rate of creation would be much too slow to be noticed. The appearance of a single new hydrogen atom would be just as hard to detect as a new sand-grain in the Sahara, so that on purely experimental grounds the steady-state idea could be neither confirmed nor disproved.

However, there was another means of attack. If the universe is in a steady state the average distance between galaxies will not alter over vast periods of time. When we look at systems thousands of millions of light-years away, we are seeing them as they used to be thousands of millions of years ago, so that in effect we are looking backward through time. Radio methods can reach out further than optical observation, and at Cambridge Sir Martin Ryle and his team found that the distribution of very remote galaxies is not the same as for galaxies closer to us. If so, the universe has changed radically, and cannot be in a steady state. The continuous creation theory was intriguing, and many attempts were made to rescue it, but it did not stand up to investigation, and by now it has been definitely cast on to the astronomical scrap-heap.

Note, however, that there is nothing irrational in the actual assumption of the continuous creation of matter. The material in the universe exists; therefore, it was created; why should not this happen in small quantities, in different areas, rather than in one place at one moment. We are back to our old difficulty, but on this score alone the one theory is as plausible as the other.

To replace the steady-state idea we have the cyclic or oscillating universe, in which a phase of expansion (such as that at the present time) will be followed by one of contraction. Eventually all the galaxies come together once more, and the whole story begins anew. We can then take our choice between a set moment of creation, or an infinite past. Everything, however, depends upon whether the average density of matter in the universe exceeds a certain critical limit. If it does, then the galaxies cannot continue to race apart indefinitely; gravitation will draw them back, and the cyclic picture will be correct, with a new "big bang" every 80,000 million years or so. If not, then nothing can half the recession. The critical value works out at the equivalent of one atom of hydrogen per 10^{23} cubic metres — about twice the volume of the Earth's globe. Present evidenve indicates that the actual density is below the critical limit, and in this case the cyclic theory must be given up; but the data are far from reliable, and the whole question remains open.

Clearly there is no overall agreement about the evolution of the universe, and there is one more point which is worth noting, because it is of fundamental importance. Suppose that the red shifts in the spectra of galaxies are not pure Doppler effects?

The Doppler effect has been tested time and time again, and there is no doubt that it is valid over distances which we can check independently. All attempts to explain the red shifts in other ways rest upon totally unproved assumptions. It has been suggested, for instance, that light is affected by interactions with material spread between the galaxies, so that its energy is lessened and its wavelength increased, but we have no proof that this is a real possibility, which is one reason why astronomers tend to be sceptical about it. Yet the red shifts in the spectra of some quasars show anomalies; and according to almost all authorities, quasars include the most remote objects known to us. In America, Halton Arp has drawn attention to quasars which are signigicantly aligned with ordinary galaxies — but have different red shifts.

If the red shifts of quasars are not due solely to the

Doppler effect, then we must also re-examine the red shifts of galaxies in general, and the whole expanding-universe theory might be in danger, making our current cosmologies as outdated as those of a century ago. Admittedly this is not likely, and it represents a minority view, but we have to admit that we cannot yet discount it.

There is, too, a revolutionary theory due to Hannes Alfven, the chief founder of the new sciency of magneto-hydrodynamics. It is as exciting as it is revolutionary, and if valid it will lead to a major overhaul of all cosmological thought. It assumes the existence of two kinds of material; ordinary matter of the type familiar to us, referred to by Alfven as koinomatter, and its exact opposite, termed antimatter. If the two types meet they will annihilate each other, releasing energy. Alfven points out that there would be no means of telling whether a galaxy were composed entirely of koinomatter or entirely of antimatter. The appearance, visually and spectroscopically, would be exactly the same, so that for all we know the Andromeda Spiral might be an antimatter system. Of course, there can be no antimatter on the Earth, and probably none in the Solar System. There might be none in our whole Galaxy, but we must always remember that the Galaxy is a very small unit by cosmical standards.

Over most parts of the universe, mixing of the two types of matter might be prevented by something roughly analogous to the Leidenfrost phenomenon. This effect was originally studied by means of the behaviour of a drop of water placed on a hot plate raised to a temperature of several hundreds of degrees. A layer of steam is formed between the drop and the hot plate, forming an insulating layer which prevents the heat from being conveyed rapidly through to the drop. Under suitable conditions the drop may persist for several minutes, though if the plate is heated less violently the steam layer is too thin to provide effective insulation, and the drop evaporates at once.

On Alfvén's theory this principle can be applied to an 'ambiplasma', or mixture of koinomatter and antimatter. The first contact between the two types will result in the mutual annihilation of particles and antiparticles, but energy will be generated, and this will result in a force capable of separating the two types until the mixing has to all intents and purposes ceased. Alfvén calculates that a cosmical Leidenfrost layer might be only 1/1000 of a light-year across, or perhaps even less – that is to say 6000 million miles, which is very slight compared with the distances between galaxies or even between individual stars.

If the universe began as a very tenuous ambiplasma, covering a huge area, there would be gravitational contraction and consequent annihilation of koinomatter and antimatter, but this would not continue indefinitely. Radiation pressure would finally become strong enough to halt the contraction, so that expansion would begin, and the two species of matter would become separated, insulated from each other by Leidenfrost layers. Probably the expansion would be indefinite; if not, then the material would come together again and there would be further extensive annihilation before the new phase of expansion.

The main obstacle to Alfvén's theory is that we have no proof of the existence of antimatter of this sort. Of course, this may not be a really valid objection, and all we can say is that the whole hypothesis rests upon an assumption which cannot be checked. (Suggestions that the Tunguska meteorite, which landed in Siberia in 1908, was a piece of antimatter are not taken seriously.) Whether there can ever be experimental confirmation or denial remains to be seen. It must be added, however, that the annihilation process is particularly suitable for explaining the intense luminosity of the quasars.

Speculations of this kind would have been out of the question half a century ago. The external nature of the galaxies was proved only in the inter-war period, so that remarkable progress has been made in a very short time. Whether we yet have anything like an accurate picture of the universe cannot be regarded as certain; most modern astronomers are confident that we are on the right track, but past generations have been equally confident, only to find themselves shown to be wrong. There may be many surprises in store for us during the next few decades, but at least we can claim that there is no longer any danger of observed facts being questioned on unscientific grounds. This in itself is a major advance, and it belongs essentially to our own age.

11. The Supreme Unimportance of Man

Mankind was once regarded as supreme. The Earth lay in the centre of the universe, with all other bodies revolving round it; the Sun, Moon and stars existed only for our benefit. Yet, rather surprisingly, this did not mean that the Earth was taken to be the only inhabited world. The existence of living beings elsewhere was more or less taken for granted, and this belief continued into relatively modern times. As we have noted, Sir William Herschel went so far as to claim that there was a habitable region below the surface of the Sun. And in this connection it is worth quoting a passage from a famous book written in the 17th century by James Ferguson.

Ferguson had a curious career. He was the son of a farm labourer, and became a shepherd-boy at the age of 10. While guarding his sheep at night, he also watched the stars; a local butler taught him mathematics, and he subsequently became famous as a maker of astronomical clocks, sun-dials and orreries. He was also the first great British popularizer of astronomy, and his book ran to many editions. When discussing comets, he wrote:

'The extreme heat, the dense atmosphere, the gross vapours, the chaotic state of the Comets, seem at first sight to indicate them altogether unfit for the purposes of animal life, and a most miserable habitation for rational beings; and therefore some (Mr Whiston, in his *Astronomical Principles of Religion*) are of the opinion that they are so many hells for tormenting the damned with perpetual vicissitudes of heat and cold. But when we consider, on the other hand, the infinite power and goodness of the Deity . . . it seems highly probable, that such numerous and large

masses of durable matter as the Comets are, however
unlike they be to our Earth, are not destitute of beings
capable of contemplating with wonder, and acknowledging
with gratitude, the wisdom, symmetry and beauty of the
Creation.'

Ferguson was evidently prepared to believe in intelligent
life which might be quite different from our own even though
it was presumably the work of the same Creator. This attitude
was not uncommon at the time, and for that matter it is not
uncommon even now, though it is hardly likely that anyone,
scientist or layman, will accept the idea of habitable comets.

It is not easy to decide just what the ancient peoples really
thought about extraterrestrial life. The surviving writings are
usually mixed in with deliberate fantasy, or in some cases –
notably Lucian's *True History*, probably the first of all
science fiction stories – made up of fantasy without any science
at all. Lucian of Samosata, a second-century Greek satirist,
wrote a splendid description of a war between the forces of the
King of the Sun and the King of the Moon, but he hastened
to add that his story was made up of nothing but lies from
beginning to end. Centuries later came the *Somnium*, by no
less a person than Johannes Kepler, which also dealt with a
lunar voyage, and which peopled the Moon with remarkable
creatures suited to their peculiar environment. The *Somnium*
was in fact a defence of the Copernican theory, written in
pseudo-fictional form. Whether Kepler actually believed in
his Moon-beings is another matter, though he may well have
done.

At that time the airless and overwhelmingly hostile character
of the Moon was not known. Accurate lunar observing may
be said to have begun in 1779, when Johann Schröter, a
German amateur, set up an observatory at Lilienthal in the
Bremen area, and worked patiently away until his equipment
was destroyed by the French invaders in 1814. Schröter
believed that the Moon might have a reasonably dense
atmosphere, and this would mean that life there could not be
ruled out, though Schröter himself made no speculations.

After his death, lunar work was carried on by two of his countrymen, Wilhelm Beer and Johann Mädler, who produced a reliable Moon-map in 1837. Unlike Schröter, they held that there could be neither atmosphere nor life on the Moon, and most contemporary astronomers agreed with them. Yet the general public was ready to believe in lunar creatures until well into the 19th century, as was shown by two separate episodes.

The first was a pure hoax, organised in 1835 by a New York newspaper reporter named Richard Locke. At that time Sir John Herschel was at the Cape of Good Hope, busy upon making observations of the southern stars which could never be seen from Europe or the United States. Locke took the opportunity to print some remarkable stories about the various weird creatures which Herschel was supposed to have seen on the Moon, and a great many people were deceived, though admittedly not for long. Then there was Franz Gruithuisen, a serious lunar observer, who was firmly convinced that some of the structures he had seen on the Moon's surface were of artificial origin. But by the 1860s the whole idea of lunar life, at least upon our pattern, had been definitely given up. Conditions there were too obviously unfavourable.

One curious theory, due to the Danish mathematician Hansen, was that the Moon's centre of gravity was not coincident with the centre of figure, so that all the lunar air and water had been drawn round to the side which is always turned away from us and which we can therefore never see. The idea was far-fetched from the outset, and few astronomers took it at all seriously. No observational disproof was really needed, but from 1959 onward it has been possible to photograph the averted side of the Moon by means of cameras carried in space-probes, and – as expected – it has been found to be just as barren and unfriendly as the side we can see. By the end of 1967 the whole of the Moon's surface had been examined in great detail, and no possible doubts about lunar life could remain.

Gradually there came a distinct branching of thought with regard to life elsewhere in the Solar System. One branch dealt with alien life-forms which could exist quite happily under conditions that we would find intolerable, while the second was concerned with life of strictly terrestrial type. The first of these branches will be discussed below, because it is part of a larger concept. The second has always been much more straightforward, because there is enough evidence to provide a basis for discussion.

Of all the planets and satellites in the Sun's system, most could be dismissed at once as being either too hot, too cold or airless. The idea of Earth-type life upon a giant planet such as Jupiter is clearly out of court, and by 1900 the only remaining candidates were the two nearest planets, Venus and Mars.

Venus is covered with a dense, cloudy atmosphere which hides its surface most effectively. As recently as 1962 very little was known about it, though the scanty information available was not encouraging. Spectroscopic work had shown that the most abundant gas in the atmosphere was carbon dioxide, but this was not conclusive, because only the uppermost layers were available for examination, and it was suggested – quite reasonably – that conditions lower down might be very different. According to F. L. Whipple and D. H. Menzel, the planet was likely to be ocean-covered, with clouds made up largely of H_2O. The surface temperature was undeniably high, but it seemed quite on the cards that there might be primitive life in the seas. Life on Earth began in our oceans, at a time when the terrestrial atmosphere contained much more carbon dioxide and much less free oxygen than it does now; why should not Venus be a world upon which life was just beginning, and would eventually develop into highly complex forms?

This was the most that could be expected. An earlier theory, due to the Swedish physicist Svante Arrhenius, had described Venus as a carboniferous-type world, with luxuriant vegetation of the primitive type found on Earth during the Coal Forest period. The detection of large amounts of carbon dioxide had indicated otherwise, but there was no valid reason to reject the basic idea of life on Venus, and this would still be the situation but for developments in space research.

The first successful planetary probe was the United States vehicle Mariner 2, which by-passed Venus in December 1962 at a distance of roughly 21,000 miles, and sent back information showing that the surface temperature really is very high. Subsequent probes have confirmed this, and have given us a picture of Venus which is, to put it mildly, unprepossessing. The surface temperature is of the order of 900 degrees Fahrenheit; the atmospheric pressure is crushing, and is about 90 times as great as that of the Earth's air at sea level; the atmosphere consists mainly of carbon dioxide, while the clouds contain large quantities of sulphuric acid. There are craters and ravines, and there are rocks strewn around in vast numbers — as we know from the Russian probes Venera 9 and Venera 10, which soft-landed on Venus in October 1975 and sent back one picture each before being put out of action by the intensely hostile conditions. By now it seems safe to say that any Earthy-type life on Venus is completely out of the question.

Mars is a different kind of world, and had always been regarded as a possible abode of life. In 1877 the Italian astonomer G.V. Schiaparelli had reported strange, straight, artificial-looking lines which he called "canali", and which became known as canals; Percival Lowell, founder of the famous observatory at Flagstaff in Arizona, firmly maintained that they were artificial waterways, constructed by intelligent beings to provide a planet-wide irrigation system upon an arid world. Lowell's views were violently criticized even in his lifetime (he died in 1961), but until the coming of the space-probes it was still generally believed that even though advanced life could not be expected, the dark tracts upon the planet were probably due to organic matter — in other words, to primitive vegetation. It was also thought that the surface was likely to be gently undulating, with no high mountain ranges anywhere.

Mariner 4, the first successful vehicle to by-pass Mars, made its approach in July 1965, and sent back pictures showing that the surface was cratered. Canals were conspicuous only by their absence. The atmosphere proved to be made up not of nitrogen, as had been believed, but of carbon dioxide, and the pressure was surprisingly low; we have now established that even on the Martian surface the pressure is below 10 millibars everywhere. In 1971-2 the even more successful Mariner 9 sent back thousands of high-quality photographs, showing huge volcanoes, deep valleys and numerous craters; the dark regions are not basically different from the ochre tracts, except in colour, and the vegetation theory had to be abandoned. Then, in 1976, two Viking probes made controlled landings, one in the region of Chryse and the other in Utopia, and carried out an intensive search for living matter. Nothing was found, and by now most authorities have come to the reluctant conclusion that Mars is a sterile world.

Frankly, this was a disappointment. If life had been found, it would have been safe to conclude that living things will appear wherever conditions are suitable for them, and will develop as far as possible in consideration of their environment. Mars is not overwhelmingly hostile, and there is plenty of evidence of past water activity; is it possible, then, that there used to be life, even if there is none today? As yet we can give no definite answer, and we must await the analysis of the first samples to be brought back to Earth from Mars. But for the moment, we have to conclude that we are probably living upon the only world in the Solar System which contains life of any sort.

Matters would be far easier if we were confident as to how life began on our own Earth, but here too we are on uncertain ground. Svante Arrhenius once suggested that life came to Earth by way of a meteorite. This "panspermia" theory never met with much support, and neither has a much more recent suggestion by F. Hoyle and C. Wickrama-singhe that the life-bearing body was a comet. But the last word has by no means been said.

To sum up: in the Solar System, at least, only the Earth seems able to support advanced life of our own kind. When we come to consider alien life-forms we are reduced to rather unprofitable speculation, because we have no concrete facts to guide us. It is impossible to disprove the existence of alien life, but there is no evidence in support of it, and all the indirect evidence which we can muster indicates that weird creatures of the science-fiction variety do not exist in the Solar System or anywhere else.

On this score there has been no marked change of view in

modern times, though it is true that of late there have been some highly speculative papers which have been treated more seriously than they would have been a few decades ago. There is no point in discussing them here, and there is even less point in doing more than mention the flying-saucer craze, which began in 1947 and which still lingers on. Regretfully, the idea that flying saucers (or Unidentified Flying Objects, to give them their more pretentious title) are visiting space-ships must be placed on a par with Richard Locke's Moon-creatures of 1835.

But though we must discount the possibilities of intelligent life elsewhere in the Solar System, we cannot do so when we come to consider the universe as a whole. Here there has been a distinct change of outlook, linked with the changing theories of the origin of planetary families.

To recapitulate: in the early part of our century the passing-star hypothesis was generally favoured, so that Solar Systems would necessarily be extremely rare. Today we think differently, and it is reasonable to assume that many stars of the same type as the Sun will develop planet families of their own. Presumably there will be planets of other stars which are similar to the Earth, and which are therefore suitable for the development of our type of life.

There are 100 000 million stars in our Galaxy. The Palomar reflector can photograph about 1000 million galaxies, so that the total number of known suns is staggeringly great, and to suppose that our Sun is unique in being attended by a habitable planet is surely illogical. It would not be likely even if the Solar System were due to a freakish encounter between two stars, and it seems to be quite out of the question if, as is now believed, planetary systems are common.

What we cannot prove, of course, is that a habitable planet is necessarily inhabited. Here we come back to speculation, though there are strong reasons for assuming that life will appear wherever conditions are suitable for it.

Rocketry can never solve this problem. To send a material probe to another star seems to be impossible either now or in the future; even at the velocity of light, which is theoretically

unattainable, a round trip would take year. The only hope
of contacting other races seems to be by radio, and experiments have already been conducted, though — predictably
— with negative results.

In 1960 a team of radio astronomers at Green Bank, in
West Virginia, 'listened out' at the 21-cm wavelength in the
hope of picking up rhythmical signals which might be of
artificial origin. This particular wavelength is that of the radio
emission of the clouds of cold hydrogen in the Galaxy, and so
radio astronomers elsewhere might be expected to concentrate
upon it. The Green Bank experiment, known as Ozma, was
discontinued after a few months, but it was at least a
serious attempt.

Even in the wildly improbable event of artificial signals
being picked up, the most we could ever hope to do would be
to establish that other intelligences did once exist in the
Galaxy. We cannot say 'do exist', because radio waves
travel at the velocity of light, and our information would
always be out of date. For instance, if signals were received
from a planet moving round a star 50 light-years away, they
would have been transmitted 50 years ago.

Here again we sense a change in outlook. If the Ozma
project had been proposed in 1930 instead of 1960, it would
have been ridiculed (even if the technical equipment had been
available then, which of course was not so). In the modern
view, life is likely to be very widespread in the universe,
even though not in our Solar System. Whether we shall ever
contact other races remains to be seen. With our present
knowledge and our present techniques the chances of success
are remarkably slight, but they are not nil.

And this, surely, brings home once more the complete
reversal in outlook between ancient and modern times. To
our ancestors, the Earth was supreme and Man unique. The
heavens existed for our benefit, and we were of paramount
importance. As knowledge grew, illusion after illusion was
swept away; first the Earth, then the Sun, then the Galaxy
became relegated to the status of a minor unit. By now we
have no illusions left on this score, and we have advanced

enough to appreciate how insignificant we really are.

We cannot tell what new information will be collected in the future, but it seems fairly safe to assume that our basic ideas will not have to be altered so drastically in the coming years as they have been in the past. We may assume, too, that elsewhere in the universe there are races whose intelligence far surpasses our own. On a peaceful Earth, mankind may in turn gain knowledge that is quite beyond our present comprehension, but this depends upon our ability to avoid any more major wars. We have to admit that the outlook in this respect is uncertain, and we cannot even tell whether the life-cycle on Earth would be able to begin again after a nuclear conflict, but the next few centuries will probably be decisive.

In a brief summary, such as has been presented here, so much has had to be left out that the story may seem to be incomplete. Yet perhaps enough has been said to show that astronomical thought has developed strikingly, even if spasmodically – and we may be sure that if humanity can learn how to live in peace, we shall eventually solve at least some of the fundamental problems of the universe.